Modern Functional Quantum Field Theory

Summing Feynman Graphs

Modern Functional Quantum Field Theory

Summing Feynman Graphs

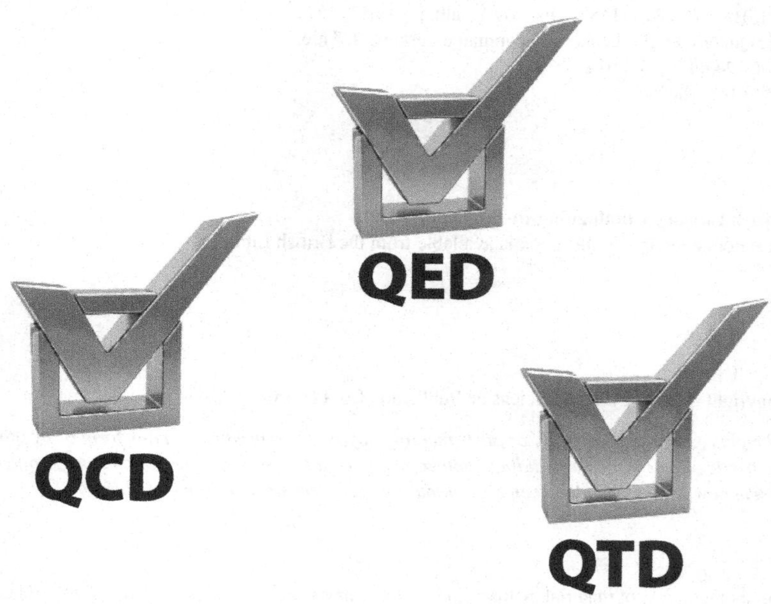

Herbert M. Fried

Brown University, USA

NEW JERSEY • LONDON • SINGAPORE • BEIJING • SHANGHAI • HONG KONG • TAIPEI • CHENNAI

Published by

World Scientific Publishing Co. Pte. Ltd.
5 Toh Tuck Link, Singapore 596224
USA office: 27 Warren Street, Suite 401-402, Hackensack, NJ 07601
UK office: 57 Shelton Street, Covent Garden, London WC2H 9HE

Library of Congress Cataloging-in-Publication Data
Fried, H. M. (Herbert Martin), author.
Modern functional quantum field theory : summing Feynman graphs / Herbert M. Fried (Brown University, USA).
pages cm
Includes bibliographical references and index.
ISBN 978-9814415873 (hadcover : alk. paper)
1. Quantum field theory. 2. Feynman diagrams. I. Title.
QC174.45.F765 2014
530.14'3--dc23
2013048545

British Library Cataloguing-in-Publication Data
A catalogue record for this book is available from the British Library.

In-house Editor: Rhaimie Wahap

Printed in Singapore

This book is dedicated to the memory of two extraordinary theoretical physicists, both for personal reasons but also in the literal sense that it could not have been written without their prior efforts: Julian Schwinger and Efimov Fradkin.

Preface

This is the author's fourth book, in a series [Fried (1972)]; [Fried (1990)]; [Fried (2002)] which has stretched over four decades, during which the applicability of Schwinger's format, functional solutions for the Generating Functional of a relativistic Quantum Field Theory have expanded from the perturbative estimates of QED radiative corrections, given by the first few orders of perturbation theory, to the non-perturbative inclusion of all relevant radiative corrections in the strong-coupling sectors of QCD.

During this period, many "partial models" have been invented, models which sum up a certain class of Feynman graphs to produce a result containing all powers of the coupling, but which leave untouched huge numbers of other graphs contributing to the same result; and therefore, such "partial models" can only be convincing in certain limiting situations. An immediate example are the eikonal "multiperipheral" or "tower-graph" models of particle scattering, which in the early '70s produced the first theoretical examples of total cross sections that satisfy the unitary requirement of the Froissart bound. As energies continue to increase, however, other unitary cancellations within the interacting structure of the omitted radiative corrections come into play, and it may be argued that[1] - within a given QFT - the total cross sections must drop below their upper, Froissart bound.

How can a sum over ALL relevant Feynman graphs describing a specific physical process be achieved? The difficulty with summing subsets of Feynman graphs is that for every increase of the order of the perturbation expansion for the desired amplitude, a new class of graphs appears and as the perturbation order increases, that new class contains an infinite number of graphs. Simply stated, summing Feynman graphs when all orders of perturbation theory are needed is a mental and physical impossibility; and

[1]See, for example, Chapter 8 of [Fried (2002)]

one must turn to a truly non-perturbative approach, one which is simple to use and powerful enough to correctly describe the complex nature of the amplitudes desired.

Happily, this approach is provided by employing (a simple variant of) the Fradkin representations for the potential-theory Green's functions and related functional, which form a basic element of (a simple variant of) the Schwinger functional solutions for every n-point function of any Quantum Field Theory (QFT). These representations allow one to perform the sum over all relevant Feynman graphs, and express the result in terms of a set of functional integrals that describe the internal, space-time fluctuations of those potential-theory Green's functions, and which can be approximated in a variety of ways, as described in the following Chapters.

For example: All previous eikonal models ever written appear trivially in the context of a Bloch-Nordsieck approximation to the exact Fradkin representation for a scattering particle.

For example: One can easily see the origins, and perform an approximate extraction, and obtain the sum of all leading UV divergences (of the inverse of the wave-function renormalization constant) of the dressed photon propagator in four-dimensional "text-book" QED; and find, upon summing the effects of an infinite set of graphs never before calculated, that the result appears to be finite. Further, in the context of this (rather good) approximation, one finds that the renormalized charge is within an order of magnitude of 1/137; and that all charged fermions obeying pure QED must have the same renormalized charge, independently of their mass.

For example: One can display a manifestly gauge — and Lorentz-invariant summation over all gluons — including cubic and quartic gluon interactions — exchanged between quarks and/or antiquarks for the calculation of all QCD n-point functions, and find that the result has the unexpected property of "Effective Locality" (EL). This property converts a main functional integral into a set of ordinary integrals; and one finds explicit demonstrations of asymptotic freedom, an analytic property no longer dependent upon any perturbative approximation.

Further, one finds that EL requires a distinction to be made between quanta of Abelian fields whose asymptotic positions or momenta can, in principle, be measured, and those of quarks which are, asymptotically, always bound, and whose transverse coordinates are, in principle, not measurable. Once this distinction between "Ideal" and "Realistic" QCD is enforced, non-perturbative absurdities of the "Ideal" theory are removed, and one can explicitly and easily obtain quark binding potentials. One small step past this realization leads, for probably the first time, to an analytic method of obtaining effective scattering and binding potentials between nucleons; and this, in turn, suggests a new way of understanding how nucleons may bind together to form nuclei.

This methodology effectively opens a new door to an understanding of the full, non-perturbative, non-linearities of QFT. It is the aim and intention of this book to state in simple, functional language this Schwinger/Fradkin approach, which has lead to the examples quoted above and described in subsequent Chapters; and to provide as much information as is presently known about sensible approximations that have been explored in this context.

For these purposes, the subject matter has been divided into four main Parts, the first of which consists of a preliminary Functional description of the techniques used; most of this preliminary material has appeared in the author's previous books, noted below, especially [Fried (1990)]. Parts II and III are devoted to QED and QCD, respectively; portions of the first Chapters of each have also been discussed, from a functional point of view, in [Fried (1972)]; and [Fried (2002)]; the new material lies in the remaining Chapters of those Parts. It should be emphasized that these techniques can be applied to any local QFT; and, indeed, to a variety of causal, non-linear problems in Physics.[2]

Part IV of this book, entitled "Astrophysical Speculations", is developed in terms of basic QFT principles, and represents the author's scientific musing on current questions of high astrophysical interest. Two basic, field-theoretic suggestions are posed, the first as the adoption of a new QED-based Model of Dark Energy; and the second as an obvious, if somewhat unusual method of extending that Model to include Inflation, which in passing provides a perfect candidate for Dark Matter and for the understanding of galactic Gamma-Ray Bursts of GeV magnitude, and of Ultra-High-Energy Cosmic Rays. One further and seemingly reasonable assumption leads to a possible description of the Big Bang, and of the Birth and Death of a Universe.

Of course, at this time there is no observational assurance for the physical correctness of these explanations, nor is any such claim intended; rather these last Chapters are meant to show how a first, and very simple field-theoretic idea can lead to another, rather revolutionary idea, which is suddenly applicable to a wide variety of astrophysical observations. The pur-

[2]See, for example, Chapter 15 of [Fried (1990)], and Chapters 6, 7, and 9 of [Fried (2002)]

pose of including this material is to show how QFT continues to contain the possibilities for a fruitful understanding of Nature, and on scales from QCD to Astrophysical.

It is the author's hope that this Schwinger/Fradkin approach will mark the beginning of a rebirth of interest in QFT, allowing it to provide measurable and verifiable predictions of experimental results, in the strong- as well as in the weak-coupling domain. In this way, QFT will pass from the adolescence of Feynman graphs, to a new maturity in which its full promise may be exhibited and appreciated.

H. M. Fried
Physics Department
Brown University

Contents

PART 1
Basic Preliminaries

Chapter 1

Quantum Field Theory – Why and How

The object of any quantum field theory (QFT) is a description of interacting particles, given in terms of a set of basic, interacting fields. The reason for this indirect approach is that physical quantities, such as energy, angular momentum, and electrical charge, conserved during the course of any interaction involving scattering, annihilation, and production of particles, are far more readily described in terms of fields than in terms of the particles themselves.

This situation arises from the admixture of quantum mechanics with special relativity: Any description of a system over a finite time interval $\triangle t$ necessarily involves an uncertainty in the energy of that system of amount $\triangle E \geqslant \hbar/\triangle t$; and since $\triangle E$ may be interpreted as $\triangle n \cdot m \cdot c^2$, where m denotes the mass of an appropriate particle, and $\triangle n$ measures the number of such particles needed in the description of that measurement, the shorter the time interval, the larger the number of virtual particles which must be included. In contrast, the field variables can be grouped into combinations whose conservation may be simply exhibited at all times, while the identification of these conserved quantities in terms of incoming and outgoing particles is performed asymptotically. These ideas were emphasized with great persuasiveness, over the last half-century, by Schwinger and by Symanzik, and they form the conceptual foundation for essentially all the material to follow.

All the formal solutions to problems of quantum and classical mechanics may be obtained by the application of a single Action Principle, first written by Schwinger [1]which generalizes other, more limited variational Principles

[1] [Schwinger (1951)]; [Schwinger (1954)]; [Schwinger (1956)]. From a slightly different point of view, much of this material may be found in his book *Particles, Sources, and Fields* [Schwinger (1970)]

known from past centuries. In particular, construction of the Generating Functionals (GF) containing all the interacting Physics of any specific theory is most simply obtained with the aid of this Action Principle, which will be exhibited below. An alternate method, subsequently introduced by Symanzik, leads to the same results using a slightly different notation.

From any GF one may, by appropriate functional differentiation, extract the form of that correlation function corresponding to the specific scattering, annihilation, or production process desired; and that form is expressed in terms of either functional integration or functional linkage operations upon a set of functionals of Potential Theory origin. Finally, the mass-shell-amputated [Symanzik (1960)] correlation functions are expressed in term of the renormalized parameters, and prediction of or comparison with experimental data is achieved.

Of course, there is many a slip 'twixt the cup and the lip, as has been said upon other occasions. The areas of specific concentration in this Book will vary, depending on the subject and the problems to which the Schwinger/Fradkin techniques have, so far, been applied: to the two-point functions in QED, and to scattering amplitudes in QCD. For the reasons explained in detail below, the techniques developed in subsequent Chapters would appear to have the greatest expectation of fulfilling the QFT tasks described in the previous paragraphs.

For these reasons, as well as for its ability to exhibit in a clear and direct manner the relevant physical structure of the particular theory under consideration, a very brief review of the Action Principle would appear to be the most efficient way to open the presentations of this book. One finds, after a few, simple manipulations involving "functional" differential equations, solutions to quantities of interest which, although formal, contain and display the structure of all the coupled equations which comprise the full theory. Perturbative approximations are always possible, if desired; but frequently, non-perturbative approximations are suggested by the formalism, and find profitable application in special physical situations.

Once one has mastered the few and very simple ideas which underly the functional game, one has a unified, non-perturbative method for treating a vast class of different physical problems.

In order to develop these topics in a sensible and coherent fashion, the following strategy will be adopted. After a brief description of the quantum mechanical underpinnings of any theory (with classical limits not discussed, but given in principle by the vanishing of Planck's constant) we shall define and briefly discuss functional differentiation and integration, with a representation of the latter by means of certain "linkage operations"; some tricks of the functional trade, as well as a useful, functional version of the cluster decomposition theorem will also be presented. After that, functional solutions to various problems of QFT will be derived for the simplest, Abelian interactions, and generalized to non-Abelian situations.

This will be followed by a brief description of Symanzik's definition of the S-matrix in terms of certain operations upon these functional solutions, so that one can proceed, in principle, from the statement of these interactions to the calculation of scattering and production amplitudes. Much of this material has already apeared in the first few Chapters of the author's previous books, with the present arrangement somewhat more convenient for the applications to follow; the latter, especially those built with the aid of Fradkin's little known but very powerful Green's function representations, forms the main thrust and content of Parts II and III of this book.

1.1 Schwinger's Action Principle

We begin, therefore, with a very brief, elementary description of Schwinger's Action Principle, one that does not nearly do justice to the beauty and clarity of the original presentations. Let the state of any quantum system be specified by an amplitude $|\mathring{a}, t>$, where $\mathring{a}$ denotes an appropriate set of all eigenvalues, discrete or continuous, which characterize the system at time t. In addition, there may be external, specified fields $J(z)$ which enter into the definition of these states via the supposed dynamics of the system; and if such "source" dependence is present the states will be denoted by $|\mathring{a}, t\rangle_J$.

Suppose that the Lagrangian of the system is called $L[Q]$, where Q denotes the appropriate variables used to describe the system. For a system with a continuously infinite number of degrees of freedom, such as a fluid or a field theory, L will be given by an integral over a Lagrangian density which depends on all spatial coordinates, $L = \int d^p x \mathcal{L}$, where $\mathcal{L} = \mathcal{L}(Q, \partial Q)$, p denotes the number of space-time dimensions and ∂ denotes a derivative with respect to time and/or spatial coordinates. We assume the Q have dynamical dependence on time; and if they are field variables, will depend on spatial coordinates as well.

There are then two essential pieces of input information which must be supplied, the Euler-Lagrange equations and the proper quantization rules. The first of these is the general equivalent of the $F = ma$ equations of motion of Newtonian mechanics, while the second describes the specifically quantum aspects of that system under consideration. It will be simplest to describe the Action Principle in mechanical terms associated with the time development of the simplest system imaginable, one containing but one degree of freedom; and at the end of the computations a straightforward generalization can be made to many-component systems. Appropriate, additional remarks will be made for systems with constraints operating between the different degrees of freedom.

The essential first step is the choice of Lagrange function (or, more simply, the Lagrangian), which contains within it all the assumed, kinematical and dynamical properties of the system. One specifies $L = L[Q, dQ/dt, t]$ and performs variations in such a way that the desired equation of motion is obtained as well as the correct (sometimes called canonical) statement of the quantum behavior of the system; both properties follow from a single Action Principle obtained by variation of the action function $W_{21} = \int dtL$, where the indices $2, 1$ refer to the time-limits of integration, from t_1 to t_2. Many different sorts of variations are possible, and useful, requiring a slight change of notation; and some of these will be considered as needed below.

The type of variations first introduced historically, and the simplest to consider, is a "parametric" variation of Q; that is, one supposes that $Q(t)$ is generalized to some integrable function $Q(t, \beta)$, and then defines the parametric variation $\delta_0 Q = \delta\beta \cdot \partial Q/\partial\beta$. One imagines variations of W_{21} due to such $\delta_0 Q$, such that $\delta_0 Q$ vanishes at the end-point times. More generally, one can at the same time include the variations induced in the action by a small change in the end-point times, $t_{1,2}$, replaced by $t_{1,2}+\delta t_{1,2}$. The corresponding total variation in W_{21} is given by

$$\delta W_{21} = \int_{t_2}^{t_1} dt \cdot \delta Q_0 \cdot \{\partial L/\partial Q - (d/dt)\,[\partial L/\partial(dQ/dt)]\}$$
$$+ \left(\partial Q_0 \cdot [\partial L/\partial(dQ/dt)] + \delta t \cdot L\right)_{21}, \tag{1.1}$$

where $A_{21} = A_2 - A_1$, and where an integration by parts has been performed on that term corresponding to the parametric variation of the dQ/dt dependence of L. The traditional Action Principle now asserts that the only nonzero variations δW are those which depend on end-time quantities; thus that first RHS term of (1.1) must vanish. With the specific application defined below, this Principle serves to specify quantum dynamics as well.

For arbitrary $\delta_0 Q$, the latter's coefficient in the time integral of δW must therefore vanish, leading to the Euler-Lagrange equation of motion,

$$\partial L/\partial Q - (d/dt)[\partial L/\partial(dQ/dt)] = 0. \tag{1.2}$$

Rewriting the remaining end-point quantities of δW in terms of the total variation of $\delta Q = \delta_0 Q + \delta t \cdot dQ/dt$ at such times, one sees that

$$\delta W = \delta Q \cdot P - \delta t \cdot H, \tag{1.3}$$

with $P = \partial L \partial(dQ/dt)$ and $H = [dQ/dt] \cdot \partial L/\partial(dQ/dt) - L$. The quantities P and H which appear in this way as coefficients of end-point coordinate- or time-variations are precisely the usual definitions of momentum and energy; remembering that all quantum changes of any system are to be effected by changes induced by unitary operators in either the operators corresponding to the system's variables or in the system's state vectors, one anticipates the identification of P and H with those quantities obtained by (1.3).

To see how this goes, consider the quantity $< \mathring{a}, t_2|\mathring{a}', t_1 >$, which for $t_1 \neq t_2$ represents the probability amplitude for the system specified by the set of quantum numbers $\mathring{a}'$ at t_1 to evolve into the state specified by quantum numbers $\mathring{a}$ at t_2. If the two times are equal, this is a transformation function, or overlap amplitude between the same or different representations of the system; that is, if the $\mathring{a}$ and the $\mathring{a}'$ are eigenvalues of completely different operators, this just represents the projection of one basis onto another. (If they are eigenvalues of the same basis, their overlap is zero if $\mathring{a} \neq \mathring{a}'$.) The differential characterization of this amplitude, the way in which it responds to small variations of various sorts, was the Action Principle discovered by Schwinger,

$$\delta < \mathring{a}, t_2|\mathring{a}', t_1 >= i < \mathring{a}, t_2|\delta W_{21}|\mathring{a}', t_1 > \tag{1.4}$$

where δW denotes not only the parametric and end-point variations discussed above but also those obtained in more general ways, e.g., by varying some parameter upon which the Lagrangian may depend.

One assumes but two properties for the description of physical systems by such state vectors, reality

$$< \mathring{a}, t_2|\mathring{a}', t_1 >^* = < \mathring{a}', t_1|\mathring{a}, t_2 > . \tag{1.5}$$

and closure

$$\sum_n < \mathring{a}, t_2|n', t >< n', t|\mathring{a}', t_1 >=< \mathring{a}, t_2|\mathring{a}', t_1 > . \tag{1.6}$$

If the expression of a general infinitesimal transformation, corresponding to the unitary operator U acting on a state vector $|å, t>$ is given as

$$\delta|å, t> = -iG(t)|å, t>, \delta < å, t| = +i < å, t|G(t), \tag{1.7}$$

with G defined by the infinitesimal operator $\delta(U - 1) = -iG$, then a comparison of (1.4) and (1.7) leads to the identification

$$\delta_{21}W = \{G(t_2) - G(t_1)\}. \tag{1.8}$$

Then, under the sort of variations leading to (1.3), one identifies

$$G(t) = P\delta Q - H\delta t \tag{1.9}$$

as the infinitesimal unitary operator for transformations appropriate to coordinate changes and to time translations. The former involves that operator, P, corresponding to changes in Q, while the latter specifies $-H$ as the operator that generates a time displacement δt; and it is these operators that must carry the fundamental commutation relations with Q and t.

The Action Principle thus specifies that the P defined by (1.3) is to be employed as the appropriate canonical momentum, with subsequent commutation relations written in terms of it, and not of some other, arbitrarily chosen quantity. To see this, imagine defining a simultaneous change in an operator $O, O' = U \cdot O \cdot U^{\dagger}$, where $U^{\dagger}$ denotes the Hermitian adjoint of U. There is then an equivalent way of writing the infinitesimal change δQ of any operator, defined as that quantity whose matrix elements reproduce the change obtained by varying the state vectors only. From (1.7) and the definition of O', one sees that δO can be written as $\delta O = i\,[O, G]$. Equating this to a c-number change δQ of the operator O itself provides exactly the desired commutation relation. For example, if $O = Q$, the appropriate generator G is given, by (1.9), as $G = P\delta Q$; and the only way to satisfy the relation $\delta Q = i\,[P\delta Q, Q]$ for c- number δQ is to adopt the commutation relation,[2]

$$[Q, P] = i. \tag{1.10}$$

If the original W is made dimensionless by dividing the RHS (1.1) by an appropriate scale e.g., $h/2\pi$ then the latter factor will just multiply the RHS

[2]If δO is to anticommute with O, as one would use for fermion fields, one is lead to the ETAR of the text.

of (1.10), and one will have precisely the fundamental statement of quantum physics (which has, as its consequence, the expression of the uncertainty principle restricting possible measurements δQ and δP of the system).

Many other applications of the Action Principle can be displayed by similar arguments, and for these the interested reader is referred to the Original Source [Schwinger (1951)]; for the purpose of the next few Chapters, the formal construction of the amplitude $< \mathring{a}, t_2 | \mathring{a}', t_1 >$, sufficient machinery is already in place. In fact, there are two different directions in which it is possible and natural to proceed. Either one specifies the kinematics and dynamics of this system by a specific choice of the Lagrangian, e.g., $L = m[dQ/dt]^2/2 - V(Q)$, in which case one is essentially treating the dynamics of a single particle in some given potential field $V(Q)$; or, one may generalize all the above discussion to systems of very many degrees of freedom - that is, to a field - and attempt to make the connection to a theory of particles and their interactions in a different, asymptotic way. It is this second method which has led to the quantum-field-theory description of relativistic particles, which possesses well-defined nonrelativistic, single-particle limits, but also contains generalizations to statistical mechanics; and it is this second method which shall be followed here.

One now imagines a system composed of very many degrees of freedom $Q_j(t) = A(x_j, t)$; and in the limit of an infinite number of such components one considers a field $A(x, t)$, with the continuous labeling of the position coordinates providing the infinite number of degrees of freedom. It is a consequence of the admixture of special relativity with quantum mechanics that the number of degrees of freedom of any theory of particles is unbounded. For measurements of a system's energy which take arbitrarily short times may have large, virtual fluctuations as described by the uncertainty principle; and those energy fluctuations δE can, in principle, be associated with the existence of an arbitrarily large number of δn of particles of mass m as given qualitatively by $\delta n \cdot mc^2$. It is one of the most attractive features of field theory that symmetries and conservation laws can be neatly described in terms of fields, while the connection with particles is made asymptotically.

All the previous analysis is now generalized to the Lagrangian description of a field, $A(x, t)$. For simplicity, this is taken to be a scalar field, rather than a spinor, vector or tensor, but appropriate generalizations of the scalar results will be appended, as necessary; by scalar, one means a field $A(x_\mu)$ transforming as a scalar under Lorentz transformations.[3] An immediate restriction is that time- and space-derivatives must enter the Lagrangian in a symmetric way. The Lagrangian L is now represented as a volume integral over all spacial dimensions of a Lagrangian density $\mathcal{L}(A, \partial A)$, symmetric in spatial and time derivatives, and the action W becomes an invariant space-time integral over the four volume lying between two space-like surfaces. It is simplest to think of such space-like surfaces as flat-time cuts, corresponding to a spatial volume at a fixed time, in which case the previous notation can be used without change, and commutation rules become equal-time commutation rules.

If source terms coupled to the fields are introduced into the Lagrangian density, e.g., a term of form $A(x) \cdot J(x)$ is added to $\mathcal{L}$, where $J(x)$ denotes some c-number scalar function, one will obtain two very convenient possibilities. Firstly, it will not be necessary to consider each amplitude $< \mathring{a}, t_2 | \mathring{a}', t_1 >$ separately, for any such quantity can be obtained by suitable choice of the $J(x)$ when computing the so-called vacuum functional $< 0, t_2 | 0, t_1 >_J$; this is, in effect, just what is done when the relation of the S-Matrix to the generating functional is prescribed in a subsequent Chapter. Secondly, the use of such source dependence allows one to display in a clear, functional manner the nonperturbative, nonlinear couplings and complexities of any nontrivial theory; one is able to see the forest as well as the trees.

1.2 Free-Field Kinematics

In the remainder of this Chapter, the kinematical properties of several standard fields will be stated, including the equal-time commutation (for bosons) or anti-commutation (for fermions) relations which follow from the Action Principle, in a manner almost identical with that simplest case sketched above; the only difference is the appearance of the additional spatial integrations, and the corresponding neglect of any spatial surface terms. In this way one finds the equation of motion for a free scalar boson field,

$$(m^2 - \partial^2)A(x) = K_x A(x) = 0, \tag{1.11}$$

[3]The relativistic notation used throughout this book is the so-called "East Coast metric", with a four-vector a_μ represented by $(\vec{a}, ia_0)$, and the product $a \cdot b = \vec{a} \cdot \vec{b} - a_0 b_0$. The Dirac matrices are Hermitian, $\gamma_\mu = \gamma_\mu^\dagger$, and satisfy the anticommutation rules $\gamma_\mu \gamma_u + \gamma_u \gamma_\mu = 2\delta_{\mu w}$, with the square of each γ_μ equal to unity.

and the equal-time commutation relations (ETCRs)

$$[A(x), A(y)]|_{x_0=y_0} = 0, \tag{1.12}$$

$$[A(x), \frac{\partial}{\partial y_0}A(y)]\Big|_{x_0=y_0} = i\delta(\vec{x} - \vec{y}) \tag{1.13}$$

which follow from the choice: $\mathcal{L}_0 = -(1/2)\{m^2A^2 + [\partial_\mu A/\partial x_\mu]^2\}$. When self-couplings of the field higher than quadratic are included in the Lagrangian, the field theory becomes nontrivial and corresponds to interactions between the particles - the quanta of this theory - via the mechanism of the exchange of an infinite number of virtual particles. As long as such nonlinear couplings do not involve terms containing derivatives of the field, the ETCRs will be the same.

If a free particle of mass m is associated with the scalar field $A(x)$, its amplitude to arrive at the spatial point x at a time x^0 if it was at y at y^0 is given by the causal propagator

$$\Delta_c(x-y) = i < (A(x)A(y))_+ > \tag{1.14}$$

with $\Delta_c(x)$ having the representations

$$\Delta_c(x) = (2\pi)^{-4}\int d^4k e^{ik\cdot x}(k^2+\mu^2-i\epsilon)^{-2}|_{\epsilon\to 0+} \tag{1.15}$$

$$\begin{aligned} &= im\theta(x^2)K_1(m\sqrt{x^2})/4\pi^2\sqrt{x^2} \\ &\quad -m\theta(-x^2)H_1^{(2)}(m\sqrt{-x^2})/8\pi\sqrt{-x^2} + \delta(x^2)/4\pi. \end{aligned} \tag{1.16}$$

Here $< ... >$ denotes a vacuum expectation value (which is always taken in these pages using an "IN" vacuum state, as described in the Chapter 3.4), and $(...)_+$ refers to a time ordering of the operators, the latest standing to the left.

Electromagnetic fields $F_{\mu\nu}$ are usually introduced in terms of the vector potential $A_\mu(x)$, but the price of this simplicity of description is the care needed in selecting the true variables of the theory, which, in turn, are dependent upon the choice of gauge. Any physical result must be independent of gauge, and so it is simplest to use the simplest formulation - the Feynman gauge - to express gauge invariant quantities, relying on the special cancellations of gauge variant quantities that follow from current conservation, or what is effectively the same thing, gauge invariance. For present purposes this means, in effect, appending an index μ and ν to the scalars A above, and multiplying the RHS of (1.13) by a factor of $\delta_{\mu\nu}$.

Useful statements for free fermion fields are the Dirac equations, with $\bar{\psi} = \psi^\dagger \cdot \gamma_4$,

$$D_x\psi(x) = (m + \gamma \cdot \partial)\psi(x) = 0, \tag{1.17}$$

$$\bar{\psi}(x)\overleftarrow{D}_x = \bar{\psi}(x)(m - \gamma \cdot \overleftarrow{\partial}) = 0, \tag{1.18}$$

and the equal-time anticommutation relations

$$\{\psi_\alpha(x), \psi^\dagger_\beta(y)\}|_{x_0=y_0} = \delta(\vec{x} - \vec{y})\delta_{\alpha\beta}, \tag{1.19}$$

$$\{\psi_\alpha(x), \psi_\beta(y)\}|_{x_0=y_0} = 0. \tag{1.20}$$

The latter pair of equations will also be assumed for interacting fermion fields. The causal propagator for a free fermion of mass m can be written as

$$S_c(x - y) = i < (\psi(x)\bar{\psi}(y))_+ > \tag{1.21}$$

$$= (m - \gamma \cdot \partial)\Delta_c(x - y; m^2), \tag{1.22}$$

where the symbol $(...)_+$ now includes an extra factor $(-1)^P$, with P representing the number of permutations of the fermion fields from the ordering displayed.

It will be useful to write a decomposition of these relativistic, free operators fields into their so-called positive and negative frequency parts. For the scalar case one has

$$A(x) = A^{(+)}(x) + A^{(-)}(x), \tag{1.23}$$

with

$$A^{(+)}(x) = (2\pi)^{-3/2} \int d^3k(2E)^{-1/2}a(\vec{k})e^{i\vec{k}\cdot\vec{r} - iEx_0},$$

$$A^{(-)}(x) = (2\pi)^{-3/2} \int d^3k(2E)^{-1/2}a^\dagger(\vec{k})e^{-ikr+iEx_0}e^{-i\vec{k}\cdot\vec{r}+iEx_0},$$

where $E = (\vec{k}^2 + m^2)^{1/2}$, and the destruction $(a(k))$ and creation $(a^\dagger(k))$ operators satisfy the commutation relations

$$[a(k), a(k')] = 0, [a(k), a^\dagger(k')] = \delta(\vec{k} - \vec{k}').$$

In a similar way, the fermion fields may be split into positive and negative frequency parts, $\psi(x) = \psi^{(+)}(x) + \psi^{(-)}(x), \psi^\dagger = \psi^{\dagger(+)} + \psi^{\dagger(-)}$, with the representations

$$\psi(x) = (2\pi)^{-3/2}\Sigma_{s=1,2} \int d^3p(m/E)^{1/2}\{b_s(\vec{p})e^{ip\cdot x}u_s(p) + d_s^\dagger(\vec{p})e^{-ip\cdot x}v_s(p)\}$$

$$\psi^\dagger(x) = (2\pi)^{-3/2}\Sigma_{s=1,2} \int d^3p(m/E)^{1/2}\{b_s^\dagger(\vec{p})e^{-ip\cdot x}u_s^\dagger(\vec{p}) + d_s(p)e^{ip\cdot x}v_s^\dagger(p)\}$$

where $E = (p^2 + m^2)^{1/2}$, and $p \cdot x = \mathbf{p} \cdot \mathbf{r} - Ex_0$. The $b_s(p)$ and $b_s^\dagger(p)$ are destruction and creation operators for fermions of spin index s and three-momentum p, while $d_s(p)$ and $d_s^\dagger(p)$ play the corresponding role for antifermions. These operators satisfy the anticommutation relations

$$\{b_s(p), b_{s'}^\dagger(p')\} = \{d_s(p), d_{s'}^\dagger(p')\} = \delta_{s,s'}\delta(\vec{p} - \vec{p}'),$$

with all other anticommutators vanishing. The Dirac spinors satisfy the normalization conditions,

$$\Sigma_{\alpha\beta} u_s^{\dagger\alpha}(p)(\gamma_4)^{\alpha\beta}u_{s'}^\beta(p) = \delta_{s,s'} = -\Sigma_{\alpha\beta} v_s^{\dagger\alpha}(p)(\gamma_4)^{\alpha\beta}v_{s'}^\beta(p),$$
$$\Sigma_{\alpha\beta} u_s^{\dagger\alpha}(p)(\gamma_4)^{\alpha\beta}u_{s'}^\beta(p) = 0$$

with the sum running over the four values of the Dirac component indices α, β for $s' \neq s$; and they also possess the closure property,

$$\gamma_4^{\alpha,\beta} = \sum_{s=1.2} \left[u_s^\alpha(p)u_s^{\dagger\beta}(p) - v_s^\alpha(p)v_s^{\dagger\beta}(p)\right].$$

The derivation of these quantities may be found in any standard text on relativistic quantum mechanics or field theory [Bjorken and Drell (1965)]; [Bogoluibov and Shirkov (1976)]; [Itzykson and Zuber (1980)], and only those few properties are quoted here which are of specific use in the arguments to follow.

From the input information of field equations plus equal time commutation or anticommutation relations, quantities which follow from different aspects of the same Action Principle, one may construct formal solutions for the time-ordered n-point Green's function of the theory,

$$< (\psi(x_1)...\psi(x_r)\bar{\psi}(y_1)...\bar{\psi}(y_m)A(z_1)...A(z_p))+ > .$$

The boson fields $A(z)$ are supposed to commute with all fermion fields at equal times; two or more kinematically independent fermions fields are assumed to anticommute at equal times. By virtue of translational invariance, each of these n-point functions is a function of $n-1$ coordinate differences. For $n=2$, one builds the propagators, the dressed or completely interacting generalizations of (1.14) and (1.22), whose behavior for large coordinate separations yields information about the excitations, or quanta, of the coupled fields. Scattering and the production of these particles are described by the amputated, mass-shell Fourier transforms of these Green's functions for $n=4$ and $n \geq 5$, respectively, while the vertex function, $n=3$, plays an important role in the evaluation of these quantities.

The most striking aspect of quantum field theory, mass and charge renormalization, along with the appearance in intermediate steps of renormalization constants Z_i, will not be a major concern of this presentation; rather, the interested reader is referred to standard and more complete works on the subject. Aside from limitations of space, the reason is that renormalization questions are, at least in Abelian theories, essentially a high-virtual-frequency, or ultraviolet affair. In non-Abelian problems, IR behavior and renormalization properties are related; but for the applications considered it will still be simplest to confine those few, necessary remarks to the appropriate Chapter.

In non-field theory contexts, and even in certain field theory applications, perhaps the most useful Green's function is that solution of the free-field equation generalized to include a specified c-number source term, or "background field" $B(z)$,

$$[m^2 - \partial^2 + B(x)]G(x,y|B) = \delta(x-y),$$

here written, for simplicity, in terms of a $B(z)$ transforming as a scalar under the Lorentz group. This type of Green's function will appear again and again in the material to follow.

Chapter 2

Functional Preliminaries

One of the most beautiful creations in modern theoretical physics is the generating functional (GF), that quantity whose functional derivatives produce all the Green's functions of an interacting field theory. Any non-trivial theory, containing interactions, can be defined by an infinite set of coupled equations relating these Green's functions, or time-ordered n-point functions. In principle, each of these n-point functions may be obtained by functional differentiation of an appropriate GF, with respect to some source function of (or conveniently introduced into) the theory. Such GFs were invented independently, and in completely different ways, by Schwinger[1] and by Feynman[2], and generalized and clarified almost immediately by Symanzik [Symanzik (1954)] and by Fradkin [Fradkin (1966)]. In particular, the special representation invented by Fradkin may be applied to any non-trivial Green's function problem, quantum mechanical or classical. In this Chapter we review and discuss some of the more useful properties associated with elementary functional operations; then, these methods will be freely used in subsequent applications.

[1]See [Schwinger (1951)]; [Schwinger (1954)], [Schwinger (1956)] and [Schwinger (1970)]. In recent years, Schwinger has stressed the utility, conceptual and practical, of a "source" approach to particles and their interactions. In the context of field theory, one might describe such "sorcery" as a framework wherein only questions pertaining to measurable, renormalized quantities are posed; and while still forcibly connected to perturbative expansions, the answers to those questions are both finite and equal to those originally found in field theory after renormalization.

[2]In contrast to the "Lagrangian" approach of Schwinger, Feynman's path to functional methods can be characterized as a "Hamiltonian" approach, described in the text coauthored with A.R. Hibbs, *Quantum Mechanics and Path Integrals*[Feynman and Hibbs (1965)]. The intuitive and special touch that was Feynman's can best be seen in his early papers on quantum field theory, [Feynman (1948)]; [Feynman (1949)], and in his book *Quantum Electrodynamics*[Feynman (1962)].

2.1 Functional Differentiation

This simplest generalization of ordinary differentiation may be defined as follows. Suppose one has a functional of $j(x)$, that is, dependence on j which can be represented as the sum of many powers of j, each multiplied by an appropriate weighting function and integrated over all coordinates,

$$F[j] = F_0 + \int F_1(u)j(u) + \frac{1}{2!}\int F_2(u_1,u_2)j(u_1)j(u_2) + \frac{1}{3!}\int F_3(u_1,u_2,u_3)j(u_1)j(u_2)j(u_3) + \cdots . \tag{2.1}$$

Here, F_0 is a constant, and the $F_n(u_1,\ldots u_n)$ are specified, symmetric functions of their variables, with integration over the entire (typically infinite) range of all variables. This supposes that $F[j]$ has a Taylor expansion in powers of j, a simplification not necessary for the general definition, but one that is more readily grasped and discussed. It is in this sense that $F[j]$ is understood to be a "functional of j". Unless otherwise noted, $\int fj$ will be used to represent an integration over all, relevant, n-dimensional space-time, $\int d^nuf(u)j(u)$.

Functional differentiation may now be defined in a manner paralleling that of ordinary differentiation. Suppose we write explicitly the coordinate u of $j(u)$ in $F[j]$; that is, u is to denote any one of variables on the RHS of (2.1). Then, one defines

$$\frac{\delta F}{\delta j(x)} \equiv \lim_{\epsilon\to 0}\frac{1}{\epsilon}\left\{F[j(u)+\epsilon\delta(x-u)] - F[j(u)]\right\}. \tag{2.2}$$

From (2.2), there follow the simple examples,

$$\frac{\delta}{\delta j(x)}\exp\left[\int fj\right] = f(x)\exp\left[\int fj\right],$$

and

$$\frac{\delta}{\delta j(x)}\exp\left[\frac{i}{2}\int j(u)f(u,v)j(v)\right] = i\int f(x,z)j(z)\cdot\exp\left[\frac{i}{2}\int jfj\right],$$

etc.

2.2 Linear Translation

There is one sort of functional differentiation operation, involving an infinite number of differentiations, which appears frequently and which can be understood in complete analogy to the similar translation operation of the ordinary calculus. That is, if

$$\exp\left[a\frac{d}{dx}\right]\cdot f(x) = f(x+a)$$

represents the ordinary translation (given by the Taylor expansion in powers of a), the functional translation of $F[j]$ is accomplished by

$$\exp\left[\int f\frac{\delta}{\delta j}\right] F[j] = F[j+f], \tag{2.3}$$

where $\int f\frac{\delta}{\delta j} = \int d^n u f(u)\frac{\delta}{\delta j(u)}$. Equation (2.3) is surely intuitive, and it may be easily proven by replacing f on the LHS of (2.3) by λf, and constructing the differential equation corresponding to variations of the parameter λ. If

$$F_\lambda[j] \equiv \exp[\lambda\int f\frac{\delta}{\delta j}]\cdot F[j],$$

then

$$\frac{\partial}{\partial\lambda}F_\lambda[j] = \int f(u)\frac{\delta}{\delta j(u)}F_\lambda[j]. \tag{2.4}$$

The general solution to (2.4) can be found in a variety of ways, but we do this here by a method which will have repeated application to other, more difficult problems. One first finds a convenient representation for an arbitrary $F[j]$, which can be given in terms of functional differentiation with respect to a different source function, $g(z)$; that is, for the functional F of (2.1) one may write

$$F[j] = \left\{F_0 + \int F_1(u)\frac{1}{i}\frac{\delta}{\delta g(u)} + \frac{1}{2!}\int F_2(u_1,u_2)\frac{1}{i}\frac{\delta}{\delta g(u_1)}\right.$$
$$\left.\cdot\frac{1}{i}\frac{\delta}{\delta g(u_s)} + \cdots\right\}\exp[i\int gj]|_{g\to 0},$$

or

$$F[j] = O\left[\frac{1}{i}\frac{\delta}{\delta g}\right] \cdot$$
$$\cdot \exp\left[i\int gj\right]\Big|_{g\to 0}, \tag{2.5}$$

where one is instructed, in (2.5), to set the source of g equal to zero after taking all necessary functional derivatives with respect to g. This is convenient for our purposes, for all the functions $F_n(u_1, ...u_n)$ are combined with functional derivatives with respect to g to form the operator $O(\delta/i\delta g)$; all the j-dependence sits in the exponential factor of (2.5), and commutes with the $\delta/\delta g$-dependence. If, therefore, one can solve (2.4) for the special choice $F[j] = \exp[i\int gj]$, for arbitrary g, then the application of $O(\delta/i\delta g)$ to that F will produce a solution of (2.4) for a general $F[j]$.

Accordingly, we rewrite (2.4) for the special functional $F_0[j] = \exp[i\int jg], F_\lambda^0[j] = \exp[\lambda\int \frac{\delta}{\delta j}] \cdot F_0[j]$, which now, clearly, satisfies the differential equation

$$\frac{\partial}{\partial\lambda}F_\lambda^0[j] = (i\int fg)F_\lambda^0[j]. \tag{2.6}$$

With the proper boundary condition at $\lambda = 0$, the solution to (2.6) is immediate,

$$F_\lambda^0[j] = \exp[i\lambda\int fg] \cdot F_0[j],$$

or

$$F_\lambda^0[j] = F_0[j+\lambda f]. \tag{2.7}$$

Operation upon both sides of (2.7) with $O(\delta/i\delta g)$ then yields the expected results, (2.3), for a general functional. An equivalent but somewhat simpler derivation may be obtained by using the idea of a functional Fourier transform, as in the discussion following (2.18), below.

2.3 Quadratic (Gaussian) Translation

There is another form of "translation", introduced by Schwinger which frequently appears, and which has relevance in other, functional integral contexts. This results from the use of a "quadratic translation" operator,

$$\exp \mathcal{D} = \exp \{ -(i/2) \int \delta/\delta j(u) A(u,v) \delta/\delta j(v) \},$$

where $A(u,v)$ is a symmetric function of its variables; and we next consider its application upon some simple forms. It turns out that $[\exp \mathcal{D}] F[j]$ corresponds to a form of Gaussian functional integration, and may be obtained in closed form only when $F[j]$ itself is not more complicated than a Gaussian.

The simplest quantity of this form is $\exp \mathcal{D} \cdot \exp[i \int gj]$, and may be evaluated by replacing A in $\mathcal{D}$, by λA, and then obtaining and solving the simple differential equation corresponding to variation of λ. One immediately finds

$$\exp\Big[-\frac{i}{2} \int \frac{\delta}{\delta j} A \frac{\delta}{\delta j} \Big] \cdot \exp\Big[i \int jg \Big] = \exp\Big[\frac{i}{2} \int gAg + i \int jg \Big]. \tag{2.8}$$

There is a simple but useful generalization of (2.8) which involves the action of $\exp \mathcal{D}$ upon a product of functionals, say $F_1[j]$ and $F_2[j]$. Again, as in (2.5), it is simplest to write each of these as $O_i(\delta/i\delta g_i) \exp[i \int g_i j]$, with the g_i vanishing after all derivatives are taken. Then, since $\exp \mathcal{D}$ commutes with the $O_i(\delta/i\delta g_i)$, one can write

$$\begin{aligned} \exp\left[-\frac{i}{2} \int \frac{\delta}{\delta j} A \frac{\delta}{\delta j} \right] \cdot F_1[j] \cdot F_2[j] &= O_1\Big[\frac{1}{i} \frac{\delta}{\delta g_a} \Big] \cdot O_2\Big[\frac{1}{i} \frac{\delta}{\delta g_2} \Big] \cdot \\ &\cdot \exp\Big[-\frac{i}{2} \int \frac{\delta}{\delta j} A \frac{\delta}{\delta j} \Big] \cdot \exp\Big[i \int j(g_1 + g_2) \Big] \Big|_{g_i \to 0}, \end{aligned}$$

which, using (2.8), can be rewritten as

$$\begin{aligned} O_1 \cdot O_2 \cdot \exp\Big[\frac{i}{2} \int g_1 A g_1 + \frac{i}{2} \int g_2 A g_2 + i \int g_1 A g_2 \Big] \cdot \\ \cdot \exp\Big[i \int j(g_1 + g_2) \Big] \Big|_{g_i \to 0}, \end{aligned}$$

or,
with (2.3) and $\mathcal{D}_i = -(i/2) \int (\delta/\delta j_i) A (\delta/\delta j_i)$, $\mathcal{D}_{12} = -i \int (\delta/\delta j_1) A (\delta/\delta j_2)$, can be written in the form

$$e^{\mathcal{D}} \cdot F_1[j] F_2[j] = e^{\mathcal{D}_{12}} [(e^{\mathcal{D}_1} F_1[j_1])(e^{\mathcal{D}_2} F_2[j_2])]|_{j_1 = j_2 = j}, \tag{2.9}$$

and can be extended to products of more than two functionals, in an obvious manner. In words: $\exp \mathcal{D}$ is a "linkage operator", which first links all pairs of j-factors within each functional by the operation of $\exp \mathcal{D}$ on that functional (these may be called the "self-linkages"); and which then links the different functionals by means of the factors $\exp[\mathcal{D}_{ij}]$. These forms will appear with a certain frequency when describing the Green's functions of quantum field theory.

The evaluation of $\exp \mathcal{D}$ upon the Gaussian functional $\exp[(i/2)\int jBj]$ is somewhat more interesting, and we here follow the parametric method of Zumino [Zumino (1958)] and of Sommerfield [Sommerfield (1963)] by considering the quantity

$$F_\lambda[j] = \exp[-\frac{i}{2}\lambda \int \frac{\delta}{\delta j} A \frac{\delta}{\delta j}] \cdot \exp[\frac{i}{2}\int jBj], \tag{2.10}$$

where both $A(u,v)$ and $B(u,v)$ are symmetric functions of their arguments. Again, one constructs the differential equation corresponding to the variation of λ,

$$\frac{\partial F_\lambda[j]}{\partial \lambda} = -\frac{i}{2}\int \frac{\delta}{\delta j(u)} A(u,v) \frac{\delta}{\delta j(v)} \cdot F_\lambda[j]. \tag{2.11}$$

An intuitive ansatz is then chosen for this Gaussian quantity,

$$F_\lambda[j] = \exp[\frac{i}{2}\int j\chi_\lambda j + i\int h_\lambda j + L_\lambda] \tag{2.12}$$

where $\chi_\lambda(u,v), h_\lambda(u)$ and L_λ are three functions of u, v, and λ to be determined. Note that these quantities satisfy the boundary conditions: $\chi_o(u,v) = B(u,v), h_o(u) = 0, L_0 = 0$.

Substituting (2.12) into (2.11), canceling a factor of $F_\lambda[j]$ from both sides of the equation, and equating coefficients of $j(z)$ of the remaining terms, one finds the simultaneous relations

$$\frac{d\chi_\lambda(u,v)}{d\lambda} = \int \chi_\lambda(u,w)A(w,z)\chi_\lambda(z,v),$$
$$\frac{dh_\lambda(u)}{d\lambda} = \int \chi_\lambda(u,w)A(w,z)h_\lambda(z),$$

and

$$\frac{dL_\lambda}{d\lambda} = \frac{1}{2}\int A(u,v)\chi_\lambda(v,u) + \frac{i}{2}\int h_\lambda(u)A(u,v)h_\lambda(v).$$

In a more compact, matrix notation, these equations read

$$\frac{d\chi_\lambda}{d\lambda} = \chi_\lambda A \chi_\lambda, \frac{dh_\lambda}{d\lambda} = \chi_\lambda A h_\lambda, \frac{dL_\lambda}{d\lambda} = \frac{1}{2} Tr[A\chi_\lambda + \frac{i}{2} h_\lambda^T A h_\lambda,$$

and can be solved making use of the conditions $\chi_0 = B$, $h_0 = 0$, $L_0 = 0$. One immediately sees that $h_\lambda(u) = 0$; and that $\chi_\lambda = B[1 - \lambda AB]^{-1}$; and that $L_\lambda = -(1/2) Tr \ln[1 - \lambda AB]$. With $\lambda = 1$, we then have the desired formula,

$$\begin{aligned} \exp\big[-\frac{i}{2}\int \frac{\delta}{\delta j}\, A\, \frac{\delta}{\delta j}\big] \cdot \exp\big[\frac{i}{2}\int jBj\big] \\ = \exp\big[\frac{i}{2}\int jB(1 - AB)^{-1}j - \frac{1}{2} Tr \ln(1 - AB)\big]. \end{aligned} \tag{2.13}$$

In this shorthand, matrix notation, $\langle x|A|y\rangle = A(x,y), \langle x|B|y\rangle = B(x,y)$, while the matrix elements $\langle x|\bar{B}_\lambda|y\rangle = \langle|B[1 - \lambda AB]^{-1}|y\rangle = \bar{B}_\lambda(x,y)$ are to satisfy the integral equations

$$\bar{B}_\lambda(x,y) = B(x,y) + \lambda \int B(x,u) A(u,v) \bar{B}_\lambda(v,y),$$

or

$$\bar{B}_\lambda(x,y) = B(x,y) + \lambda \int \bar{B}_\lambda(x,u) A(u,v) B(v,y).$$

In terms of this function, $L_\lambda = \frac{1}{2}\int_0^\lambda d\lambda' \int A(u,v) \bar{B}_{\lambda'}(v,u)$.

We will shortly see that (2.13) is a statement of Gaussian functional integration.

An immediate and useful generalization of (2.13) can be seen at once:

$$\begin{aligned} \exp\big[-\frac{i}{2}\int \frac{\delta}{\delta j} A \frac{\delta}{\delta j}\big] \cdot \exp\big[\frac{i}{2}\int jBj + i\int fj\big] \\ = \exp\big[\frac{i}{2}\int jB(1 - AB)^{-1}j + i\int f(1 - BA)^{-1}j \\ + \frac{i}{2}\int fA(1 - BA)^{-1}f - \frac{1}{2} Tr \ln(1 - AB)\big]. \end{aligned} \tag{2.14}$$

In this case, the $h_\lambda(u)$ term of the ansatz (2.12) turns out to be non-zero and easily calculable.

A slightly different form of (2.13) is useful in charged Boson situations,

$$\exp\Big[-i\int \frac{\delta}{\delta j}A\frac{\delta}{\delta j^*}\Big]\exp\Big[i\int j^*Bj\Big]$$
$$= \exp\Big[i\int j^*B(1-AB)^{-1}j - Tr\ln(1-AB)\Big], \qquad (2.15)$$

and is easily obtained by the same parametric technique. Note that $A(u,v)$ and $B(u,v)$ are here no longer symmetric functions of their variables.

There exists a simple extension of (2.13) which is of immediate use in field theories containing fermions. If$A_{\alpha\beta}(u,v)$ and $B_{\alpha\beta}(u,v)$ are now non-symmetric, (Dirac) matrix-valued functions, and if the c-number fermionic sources $\eta_\alpha(u), \bar\eta_\beta(v)$ which replace the boson sources $j(u)$ are taken to be anti-commuting (Grassmann) objects, satisfying

$$\{\eta_\alpha(u),\eta_\beta(v)\} = \{\eta_\alpha(u),\bar\eta_\beta(v)\} = 0,$$
$$\left\{\frac{\delta}{\delta\eta_\alpha(u)},\eta_\beta(v)\right\} = \left\{\frac{\delta}{\delta\bar\eta_\alpha(u)},\bar\eta_\beta(v)\right\} = \delta_{\alpha\beta}\delta(u-v),$$

then a very similar construction may be carried through, yielding

$$\exp\Big[-i\int \frac{\delta}{\delta\eta_\alpha}A_{\alpha\beta}\frac{\delta}{\delta\bar\eta_\beta}\Big]\cdot\exp\Big[i\int \bar\eta_\gamma B_{\gamma\delta}\eta_\delta\Big]$$
$$= \exp\Big[i\int \bar\eta B(1+AB)^{-1}\eta + Tr\ln(1+AB)\Big]. \qquad (2.16)$$

Dirac matrices have been suppressed on the RHS of (2.16), but it should be noted that the trace operation Tr here includes a summation over Dirac indices as well as over space-time coordinates.

Note also that factors of (1/2) are absent, and that sign of A on the RHS of (2.16) appears to be reversed in comparison with the bosonic case of (2.16). More importantly, the sign of $Tr\ln[1+AB]$ is reversed, which is the origin of the "extra minus sign multiplying each fermionic closed loop" rule[3] of perturbation theory, and which plays a role in the "supersymmetric" cancellations between boson and fermion loops [Zumino (1975)].

---

[3]See, for example, any of these texts [Bjorken and Drell (1965)]; [Bogoluibov and Shirkov (1976)]; and [Itzykson and Zuber (1980)].

2.4 Functional Integration

The object here is to generalize ordinary integration over a real variable, $\int dx$, to integration over a real function $\phi(x)$. One first must define the integration measure, which is typically done by breaking up all space-time into a fine mesh of N small cells of volume $\triangle$, each labeled by a subscript$_i$, $\phi(x) \to \phi(x_i) = \phi_i$. Then, the functional integral (FI) $\int d[\phi]$ is defined as the product over all N ordinary integrals of the ϕ_i,

$$\int d[\phi] = \lim_{N\to\infty} \prod_{i=1}^{N} \int_{-\infty}^{+\infty} d\phi_i. \tag{2.17}$$

Sometimes a divergent normalization factor is included in this definition; but it really is not necessary as long as proper care is taken to normalize all physical expressions.

The simplest, non-trivial integrand to insert under the FI is one which leads to a Dirac δ-functional, expressing the equality of one function with another for all values of their arguments. Following the spirit of (2.17), one writes $\int du\phi(u)[j(u) - f(u)]$ as $\triangle \sum_{i=1}^{N} \phi_i[j_i - f_i]$ and calculates

$$\delta[j - f] = \mathcal{N}^{-1} \int d[\phi] \exp[i \int \phi(j - f)], \tag{2.18}$$

with $\mathcal{N}$ a normalization constant to be obtained. Equation (2.18) represents the δ-functional, defined in this way in terms of the product of N ordinary δ-functions, one at each space-time coordinate i,

$$\mathcal{N}^{-1} \prod_{i=1}^{N} \int_{-\infty}^{+\infty} d\phi_i \exp[i\triangle\phi(j_i - f_i)] = \mathcal{N}^{-1} (\frac{2\pi}{\triangle})^N \prod_{i=1}^{N} \delta(j_i - f_i).$$

With the normalization factor $\mathcal{N}$ chosen as $\lim_{N\to\infty} \prod_i (2\pi/\triangle)$, this $\delta[j-f]$ will under subsequent functional integration act to replace each $j(u)$ by the function $f(u)$.

Equation (2.18) is a special case of a functional Fourier transform (FFT), in the sense that the FFT of $\delta[g]$ is constant. It is frequently useful to imagine an arbitrary functional $F[j]$ as given by its FFT,

$$F[j] = \mathcal{N}^{-1} \int d[\phi] \tilde{F}[\phi] \exp[i \int j\phi],$$

and we have already done the equivalent of this in constructing the solution to (2.4), above. Again breaking up the space-time region into a fine mesh of small volumes Δ, the existence of the FFT can be understood in terms of the existence of an ordinary Fourier transform at each mesh coordinate.

It may also be noted that the measure of the FI can be defined in terms of the Fourier transform $\tilde{\phi}(k)$ of $\phi(x)$, by breaking up all of k-space into a fine mesh, and integrating the $\tilde{\phi}(k)$ in each mesh volume. However, if $\phi(x)$ is real, $\tilde{\phi}(k)$ is complex, and one must be careful to integrate over both the real and imaginary parts of each mesh variable $\tilde{\phi}_i = \tilde{\phi}(k_i)$.

Just as for the case of ordinary integration, the most complicated FI that can be performed exactly is a Gaussian. One requires

$$I[j;A] = \int d[\phi] \exp[\frac{i}{2}\int \phi A \phi + i\int j\phi], \tag{2.19}$$

where, again, $A(x,y)$ is a symmetric function of its variables and $j(x)$ is an arbitrary source function. Without performing any calculation at all one can obtain the j-dependence of (2.19), by making the variable change $\phi(x) = \chi(x) - \int dy A^{-1}(x,y)j(y)$, where the quantity A^{-1} is assumed to exist, and to satisfy the relations

$$\int du A(x,u)A^{-1}(u,y) = \int du A^{-1}(x,u)A(u,y) = \delta(x-y).$$

The exponential of (2.17) then becomes

$$\frac{i}{2}\int \chi A \chi - \frac{i}{2}\int j(x)A^{-1}(x,y)j(u),$$

so that $I[j;A] \sim \exp\left[-\frac{i}{2}\int jA^{-1}j\right]$. The proportionality constant, and in particular its A-dependence, is a quantity of considerable interest, and the simplest way to obtain it is by direct integration.

Imagine an orthogonal matrix $\langle x|M|y\rangle = M(x,y)$ with a continuous number of indices, satisfying the normalization condition

$$\int du M(x,u)M^T(u,y) = \delta(x-y), \tag{2.20}$$

with $M^T(x,y) = M(y,x) = M^{-1}(x,y)$. When configuration space is broken up into small cells of 4-volume Δ, there will be correspondingly many discrete components of this matrix, M_{ij}; and with the replacement $\delta(x-y) \to \delta_{ij}/\Delta$, one finds the discrete version of (2.20),

$$\sum_{\ell} M_{i\ell} M_{\ell j}^T = \delta_{ij}/\Delta^2. \tag{2.21}$$

The orthogonal M may be chosen such that it diagonalizes A,

$$\int M^T(x,u)A(u,v)M(v,y) = \delta(x-y)a(x), \tag{2.22}$$

where $a(x)$ represents the continuous-valued eigenvalue of A. [Note that for $A(x,y) = \delta(x-y)$, (2.22) reduces to (2.20) if $a = 1$.] For discrete indices this becomes

$$\sum_{\alpha\beta} M_{i\alpha}^T A_{\alpha\beta} M_{\beta j} = \delta_{ij} a_i/\Delta^3. \tag{2.23}$$

Indeed, for any $f(A)$ (expressable as an infinite polynomial in A), these become

$$\int M^T(x,u)\langle u|f(A)|v\rangle M(v,y) = \delta(x-y)f(a(x)), \tag{2.24}$$

and

$$\sum_{\alpha\beta} M_{i\alpha}^T f_{\alpha\beta}(A) M_{\beta j} = \delta_{ij} f(a_i)/\Delta^3, \tag{2.25}$$

where $\langle x|f(A)|y\rangle$ and $f_{ij}(A)$ denote projections of $f(A)$ in the continuous and discrete cases, respectively.

Under the variable change: $\phi(x) = \int du M(x,u)q(u)$, the FI of (2.19) becomes

$$I[j;A] = \int d[q] \exp[\frac{i}{2}q^2(x)a(x) + i\int q(x)J(x)], \tag{2.26}$$

where the "new source" $J(x)$ is related to $j(x)$ by: $J(x) = \int M^T(x,u)j(u)$. Note that (2.21) implies that the new, discrete measure $d[q]$ is exactly the same as the original $d[\phi]$, since the NXN matrix ΔM_{ij} is orthogonal, with determinant unity. The discrete version of (2.26) is

$$\prod_{i=1}^{N} \int_{-\infty}^{+\infty} dq_i \exp[\frac{i}{2}\Delta a_i q_i^2 + i\Delta q_i J_i] = \prod_i \left(\frac{2\pi i}{\Delta a_i}\right)^{1/2} e^{-\frac{1}{2}\Delta J_i^2/a_i}. \tag{2.27}$$

With the $f(A)$ of (2.24) and (2.25) given by $f(A) = A^{-1}$, the product over all cells of the exponential factor on the RHS of (2.27) yields

$$\exp[-\frac{i}{2}\int j(u)A^{-1}(u,v)j(v)],$$

which is just that found above without any integration. With $f(A)$ now chosen as $f = \ln(A)$, the A-dependence of the factors

$$\prod_{i=1}^{N}[a_i]^{-1/2} = [\det a]^{-1/2} = \exp[-\frac{1}{2}\sum_i \ln a_i]$$

can be written in terms of: $\int dx\langle x|\ln(A)|x\rangle = Tr\ln(A)$, as

$$\lim_{N\to\infty}\prod_i [a_i]^{-1/2} = \exp[-\frac{1}{2}Tr\ln A],$$

after passing to the continuous limit. Denoting the remaining factors $\prod_i(2\pi i/\triangle)^{1/2}$ by the constant C, one has the precise evaluation,

$$I[j;A] = C\exp[-\frac{i}{2}\int jA^{-1}j - \frac{1}{2}Tr\ln A]. \tag{2.28}$$

Every physically significant computation of a FI such as (2.19) is always phrased in such a way that the divergent, normalization constant C disappears from the final result.

There is another way of evaluating (2.19) which brings out the equivalence of the parametrically-obtained relation (2.13) and Gaussian functional integration. From the definition (2.19) and the translation property (2.3), there follows

$$\exp\Big[-\frac{i}{2}\int\frac{\delta}{\delta j}D\frac{\delta}{\delta j}\Big]I[j;A] = I[j;A+D]. \tag{2.29}$$

If one substitutes into (2.29) the ansatz

$$I[j;A] = N[A]\cdot\exp\Big[-\frac{i}{2}\int jA^{-1}j\Big], \tag{2.30}$$

there results

$$N[A]\exp[-\frac{1}{2}Tr\ln(1+DA)^{-1}] = N[A+D], \tag{2.31}$$

where a common factor $\exp[-i/2\int j(A+D)^{-1}j]$ has cancelled from both sides of (2.31). The latter can be rewritten in the form

$$N[A]\exp\left[\frac{1}{2}Tr\ln A\right] = N[A+D]\exp[\frac{1}{2}Tr\ln(A+D)], \tag{2.32}$$

and since the LHS depends only on A, while the RHS depends only on $A+D$, both sides must be constant, independent of A and D, from which follows

$$N[A] = C\cdot\exp\left[-\frac{1}{2}Tr\ln A\right]. \tag{2.33}$$

With C chosen as the constant of (2.28), (2.33) and (2.30) again reproduce (2.28).

The corresponding, direct functional integration over fermion variables, equivalent to (2.16), is left as an exercise for the interested reader [Berezin (1961)]. It is, on the one hand, less intuitive than that of the boson case, because of the presence of anti-commuting variables; but, on the other hand, it is somewhat simpler because all products of like Grassmannian factors vanish. Equation (2.16) plays an essential role in all field theories containing fermions, and will find frequent application in this book.

2.5 Cluster Decomposition

Consider a functional $L[A]$ which is acted upon by the linkage operator $\exp[\mathcal{D}]$, with, now, $\mathcal{D} = -(i/2)\int\delta/\delta A(u)\triangle_c(u-v)\delta/\delta A(v)$. (The notation is suggestive of a specific problem in quantum field theory, but the discussion is quite general.) For simplicity, we denote this operation upon $L[A]$ by $\bar{L}[A]$,

$$\bar{L}[A] = e^{\mathcal{D}}L[A]. \tag{2.34}$$

The question posed here is how to represent the operation

$$S[A] = e^{\mathcal{D}}\cdot e^{L[A]}. \tag{2.35}$$

One method of approach is to represent $\exp[\mathcal{D}]$ by a Gaussian FI, since from (2.19) and (2.28), with $j(x)$ replaced by $(1/i)\delta/\delta A(x)$ and $A^{-1}(x,y)$ replaced by $-\triangle_c(x-y)$, one can write

$$\exp\left[-\frac{i}{2}\int\frac{\delta}{\delta A}\triangle_c\frac{\delta}{\delta A}\right]$$
$$= \mathcal{N}\int d[\phi]\exp\left[-\frac{i}{2}\int\phi\triangle_c^{-1}\phi\right]\cdot\exp\left[\int\phi\frac{\delta}{\delta A}\right], \quad (2.36)$$

with $\mathcal{N} = (C^*)^{-1}\exp[-(1/2)Tr\ln(\triangle_c)]$. Substituting this form into (2.35) produces

$$S[A] = \mathcal{N}\int d[\phi]\exp[-\frac{i}{2}\int\phi\triangle_c^{-1}\phi + L[A+\phi]], \quad (2.37)$$

which is the FI representation equivalent to (2.35). An alternate form follows from the variable change $\chi = \phi + A$, leading to

$$S[A] = \exp\left[-\frac{i}{2}\int J\triangle_c J\right]\cdot\mathcal{N}\int d[\chi]$$
$$\cdot\exp\left[-\frac{i}{2}\int\chi\triangle_c^{-1}\chi + i\int J\chi + L[\chi]\right],$$
$$J \equiv \int\triangle_c^{-1}A. \quad (2.38)$$

The FI appearing on the RHS of (2.38) has the form of the generating functional in interacting scalar field theories, and in appropriate situations may be approximated in a semi-classical way.

There is another method of approach, with origins in the cluster expansions of statistical physics, which may sometimes be useful, especially when the functional $L[A]$ is non-local and sufficiently complicated to make approximation schemes based on the successive approximations of a FI somewhat awkward. This method converts $S[A]$ directly to an exponential whose argument is an infinite sum over a set of "connected" quantities. The representation we now derive is

$$S[A] = \exp\left[\sum_{N=1}^{\infty} Q_N/N!\right], \quad (2.39)$$

where the $Q_N[A]$ are connected functionals, initially defined as

$$
\begin{aligned}
Q_N[A] &= e^{\mathcal{D}} L^N(A)|_{conn} \\
&= \prod_{i>j=1}^{N} e^{\mathcal{D}_{ij}} \cdot \prod_{i=1}^{N} e^{\mathcal{D}_i} L[A_i]|_{conn, A_i = A}, \\
&= \prod_{i>j}^{N} e^{\mathcal{D}_{ij}} \cdot \prod_{i=1}^{N} \bar{L}[A_i]|_{conn}, A_i = A.
\end{aligned}
\tag{2.40}
$$

In obtaining (2.40), the trivial generalization of (2.9) to N factors of $L[A]$ has been written, using individual fields A_i (and followed by the final limit wherein all A_i are replaced by the same A), and the obvious notation: $\mathcal{D}_i = -(1/2) \int \delta/\delta A_i \Delta_c \delta/\delta A_i$, $\mathcal{D}_{ij} = -i \int \delta/\delta A_i \Delta_c \delta/\delta A_j$. The subscript "conn" indicates that at least one linkage must be retained between any and each pair of $L[A_k]$ terms; all terms without such linkages must be discarded. For $N = 1$, (2.40) defines $Q_1[A] = \bar{L}[A]$.

We first give a combinatoric derivation of (2.39), and then a simple, parametric construction of the Q_N. In any array of N such $Q_1 = \bar{L}$ factors, one factor-pairs in all possible ways, grouping $m_1 Q_1$s singly, $2m_2 Q_1$s in pairs to form m_2 terms Q_2, $3m_3 Q_1 s$ to form $m_3 Q_3 s$, etc. The number of ways of dividing up N such factors into $Q_1, Q_2, ...Q_n$ terms is $N![m_1!(2m_2)!(3m_3)!...(nm_n)!]^{-1}$, while the number of Q_2 pairs made is $(2m_2)!/m_2!(2!)^{[m_2]}$, the number of Q_3s formed is $(3m_3)!/m_3!(3!)^{[m_3]}$, etc. Hence the contribution to the sum made by the Nth term, where $N = m_1 + 2m_2 + 3m_3 + ... + nm_n$, is given by

$$
\frac{1}{N!} \cdot \frac{N!}{m_1!(2m_2)! \dots (nm_n)!} \cdot \left(\frac{(2m_2)!}{m_2!(2!)^{m_2}} \right)
\cdots \left(\frac{(nm_n)!}{m_n!(n!)^{m_n}} \right) \cdot Q_i^{m_1} \cdot Q_2^{m_2} \cdots Q_n^{m_n}.
$$

Since the sum over N runs from 1 to ∞, one may sum over each m_n separately to obtain (2.39).

Explicit forms of the Q_N are most easily found by considering how such quantities would change if each $L[A]$ was scaled by a constant factor λ; clearly, if $L \to \lambda L$, then $Q_N \to \lambda^N Q_N$, and $S[A]$ becomes, with (2.39),

$$
\exp\left[\sum_{n=1}^{\infty} \lambda^n Q_n / n! \right] = e^{\mathcal{D}} \cdot e^{\lambda L},
$$

or

$$\sum_{n=1}^{\infty} \lambda^n Q_n/n! = \ln[e^{\mathcal{D}} \cdot e^{\lambda L}]. \tag{2.41}$$

Each Q_N can now be obtained by calculating $(\partial/\partial\lambda)^N|_{\lambda=0}$ on (2.41). For example

$$Q_1 = [e^{\mathcal{D}} \cdot Le^{\lambda L}/[e^{\mathcal{D}} \cdot e^{\lambda L}]|_{\lambda=0} = e^{\mathcal{D}} \cdot L = \bar{L};$$

and

$$Q_2 = e^{\mathcal{D}} \cdot L^2 - (e^{\mathcal{D}} L)^2 = [e^{\mathcal{D}_{12}} - 1]\bar{L}[A_1]\bar{L}[A_2]|_{A_1=A_2=A},$$

which gives expression to the "conn" subscript of (2.40). It is somewhat easier to represent pictorially the higher Q_N obtained from (2.41), and this can be found in the corresponding Chapter of *Functional Methods and Eikonal Models* [Fried (1990)].

2.6 Two Useful Relations

There is a simple rearrangement that can be used to rewrite subsequent formal functional solutions in a more transparent manner. Suppose that $F[A]$ is a functional given by any sum of polynomials, most simply represented (as in the discussion following (2.5)) by $\exp[i \int fA]$, with $f(u)$ arbitrary and susceptible to functional differentiation. Then, it immediately follows that

$$\begin{aligned} F\left[\frac{1}{i}\frac{\delta}{\delta j}\right] &\cdot \exp\left[\frac{i}{2}\int j\mathcal{D}j\right] \\ &= \exp\left[\frac{i}{2}\int j\mathcal{D}j\right] \cdot \exp\left[-\frac{i}{2}\int \frac{\delta}{\delta A}\mathcal{D}\frac{\delta}{\delta A}\right] \cdot F[A], \end{aligned} \tag{2.42}$$

with $A(x) = \int \mathcal{D}(x,y)j(y)dy$. Equation (2.42) will be useful in converting a formal solution involving the operation of a complicated functional differential operator upon Gaussian j-dependence, to a Gaussian translation or linkage operator acting upon the same functional, $F[A]$.

A second relation which is both simple and frequently quite useful, concerns "equivalent functional variations". Again, let $\exp[i \int gA]$ be used to represent an arbitrary functional $F[A]$ susceptible to functional differentiation with respect to $g(z)$. Then, with one integration by parts, it follows that

$$\frac{\delta}{\delta\Lambda(z)}F[A+\partial\Lambda] = -\partial^z_\mu \frac{\delta}{\delta A_\mu(z)}F[A+\partial\lambda], \tag{2.43}$$

for arbitrary $\Lambda(z)$. This is used, for example, in discussing the gauge structure of Green's functions in QED.

Chapter 3

Functional Field Theory

The object of this Chapter is to present a very brief sketch of the formal, functional solution for the n-point Green's functions of local, quantum field theory, which follow from the assumptions of the Action Principle. In essence, this treatment will use Schwinger's construction [Schwinger (1951)] of the generating functional, and will illustrate the care that must be taken when discussing products of fields as the same space-time point; some very brief comments describing the independent formulation[1] of Symanzik and Fradkin will also be made. These functional solutions will be rephrased in terms of functional integrals, while some discussion will be given of the generalizations needed for non-Abelian problems. Finally, a brief discussion will be given of the singular nature of certain n-point functions at overlapping space-time coordinates, and of the role of charge conservation in alleviating that situation.

3.1 The Generating Functional

Consider the vacuum-to-vacuum amplitude in a theory of a Hermitian, scalar field $A(x)$, defined by the Lagrangian density $\mathcal{L} = \mathcal{L}_0 + \mathcal{L}'$ with the "free", or quadratic $\mathcal{L}_0$ and the interaction $\mathcal{L}'$ given by

$$\mathcal{L}_0 = -\frac{1}{2}\left[m^2 A^2 + (\partial_\mu A)^2\right], \mathcal{L}' = -gA^n. \tag{3.1}$$

A consistent theory is presumably possible only [Baym (1960)] for even n, while renormalization effects depend on the number[2] of space-time dimen-

[1]More details may be found in [Fried (1972)]

[2]In four dimensions, for $n > 4$ and with the usual definitions of renormalization, the non-perturbative theory corresponding to an A^n interaction is a free-field theory. This is believed to be the case for $n = 4$ as well. See, e.g., [Symanzik (1979)]; and the many

sions; the constants m and g here denote the bare mass and charge of the theory, and frequently are written with a subscript $_0$ to distinguish them from their renormalized counterparts. The form of $\mathcal{L}'$ can be more complicated than that of (3.1), as long as it is local; or, in a non-relativistic theory, it can be spatially non-local, but must be local in time. One assumes for simplicity that the ground state of the theory is the same as the vacuum state, the state with zero particle number; for many-particle systems this need not be true.

To this Lagrangian, it is convenient to add a source term, $\mathcal{L} \to \mathcal{L} + J(x)A(x)$ where $J(x)$ denotes a c-number, "source" field; now the entire theory, the operators $A(x)$, the state vectors, and all matrix elements depend on J. In particular, the probability amplitude to find the system in the vacuum state at a time t_2, if it was known to be described by the vacuum state at t_1, is denoted by $\langle 0, t_2|0, t_1\rangle_J$. It is this quantity which is at the heart of all formal, functional developments, a situation which remains true when generalizations of this analysis are made to other boson (e.g., vector) and fermion fields.

The first question to be posed is the explicit form of the J-dependence of this vacuum amplitude, and for this the Action Principle provides an immediate answer. For one learns from (1.4) that a variation of this amplitude due to an infinitesimal change in J is just given by

$$[\delta/\delta J(x)]\langle 0, t_2|0, t_1\rangle_J = i\langle 0, t_2|[\delta/\delta J(x)]W_{2,1}|0, t_1\rangle_J. \tag{3.2}$$

In calculating this variation of the action $W_{2,1} = \int d^4x\mathcal{L}$, one must consider both the explicit J dependence of $\mathcal{L}$ as well as the implicit J dependence of the operators $A(x)$; but because the Euler equations are to hold, as derived from parametric variations of $A(x)$, the coefficient of the implicit variations vanishes, and one is left with only the simple, explicit dependence,

$$\frac{1}{i}[\delta/\delta J(x)]\langle 0, t_2|0, t_1\rangle_J = \langle 0, t_2|A(x)|0, t_1\rangle_J, \tag{3.3}$$

where it is understood that the time coordinate x_0 must lie between t_1 and t_2; otherwise, the RHS of (3.3) is zero.

One next asks for the result when a second functional derivative is taken, at some other space-time point y. Fr this, one uses the closure property (1.6) to obtain for $t_2 > x_0 > y_0 > t_1$

references therein. In the context of an IR, non-perturbative approach, this can be seen in [Fried (1980)].

$$\left(\frac{1}{i}\right)^2 [\delta/\delta J(x)][\delta/\delta J(y)\langle 0,t_2|0,t_1\rangle_J = \langle 0,t_2|A(x)A(y)|0,t_1\rangle_J. \quad (3.4)$$

For the other time sequence, $t_2 > y_0 > x_0 > t_1$, one finds the same form as (3.4) but with the RHS $x-$ and y-dependence interchanged. The general result is then

$$\left(\frac{1}{i}\right)^2 [\delta/\delta J(x)][\delta/\delta J(y)]\langle 0,t_2|0,t_1\rangle_J = \langle 0,t_2|(A(x)A(y))_+|0,t_1\rangle_J, \quad (3.5)$$

with the ordered bracket making its natural appearance on the RHS of (3.5). Clearly, for n functional derivatives, one has

$$\left(\frac{1}{i}\right)^n [\delta/\delta J(x_1)] \ldots [\delta/\delta J(x_n)]\langle 0,t_2|0,t_1\rangle_J$$
$$= n\langle 0,t_1|(A(x_1)\ldots A(x_n))_+|0,t_1\rangle_J, \quad (3.6)$$

where the bracket of (3.6) contains n operators, ordered so that those containing the largest time-coordinates stand sequentially to the left; the RHS of (3.6) is non-zero only when all the time coordinates lie in the interval between t_2 and t_1.

By the simple expedient of setting $J = 0$ in (3.6), one constructs the coefficient of the nth power of J, in the Taylor expansion of this J-dependent vacuum amplitude; and one infers that it can be written in the form

$$\langle 0,t_2|0,t_1\rangle_J = \langle 0,t_2|\left(\exp\left[i\int d^4x J(x)A(x)\right]\right)_+|0,t_1\rangle, \quad (3.7)$$

where the states and the field $A(x)$ on the RHS of (3.7) are independent of J, and where its $\int dx_0$ runs between t_2 and t_1. Following Symanzik, one introduces the unitary operator

$$\mathcal{T}^{\overset{2}{1}}[J] = \left(\exp\left[i\int_{t_1}^{t_2} d^4x J(x)A(x)\right]\right)_+, \quad (3.8)$$

which quantity is central to his construction of the S-matrix, as presented in Chapter 4. In the $t_2 \to \infty$, $t_1 \to -\infty$ limit, the vacuum expectation value (VEV) of (3.8) defines the generating functional (GF) $Z[J]$. Because the vacuum states used for $Z[J]$ (typically, an "IN" state vacuum) are at the same time, and those appearing in (3.7) are at different times, there will appear a constant factor of proportionality relating the two functionals; but the normalization of $Z[J]$ is always chosen so that $Z[0] = 1$.

It will first be convenient to obtain the formal, functional solution for $\langle 0, t_2 | 0, t_1 \rangle_J$, by returning to (3.3) and writing the differential equation

$$K_x \left(\frac{1}{i}\right) [\ \delta\ /\delta J(x)] \langle 0, t_2 | 0, t_1 \rangle_J = \langle 0, t_2 | \big(J(x) - ng[A(x)]^{n-1}\big) | 0, t_1 \rangle_J$$

$$= J(x) - ng \left(\frac{1}{i}\right)^{n-1} [\delta/\delta J(x)]^{n-1} \} \langle 0, t_2 | 0, t_1 \rangle_J, \qquad (3.9)$$

in which the Euler equation for $A(x)$ has been used, as well as the equal-coordinate limit of (3.6). A solution for the vacuum amplitude may be constructed by considering the free-field limit, setting $g = 0$ in (3.9), and denoting the corresponding amplitude by $\langle 0, t_2 | 0, t_1 \rangle_J^0$. One finds

$$K_x [\delta/\delta J(x)] \langle 0, t_2 | 0, t_1 \rangle_J^0 = i J(x) \langle 0, t_2 | 0, t_1 \rangle_J^0,$$

with solution

$$\langle 0, t_2 | 0, t_1 \rangle_J^0 = \exp \left[\left(\frac{i}{2}\right) \int d^4x \int d^4y J(x) \Delta_c(x - y; m^2) J(y) \right], \qquad (3.10)$$

where the integration of (3.10) runs over all spatial points and all times lying between t_1 and t_2.

The simplest way of constructing a solution to the full (3.9) is to return to the Action Principle and consider variations of the coupling, $g \to g + dg$,

$$[\delta/\delta g] \langle 0, t_2 | 0, t_1 \rangle_J = -i \langle 0, t_2 | \int_{t_1}^{t_2} d^4x [A(x)]^n | 0, t_1 \rangle_J$$

or

$$[\delta/\delta g] \langle 0, t_2 | 0, t_1 \rangle_J = -i \int_{t_1}^{t_2} d^4x \left[\left(\frac{1}{i}\right) \delta/\delta J(x) \right]^n \langle 0, t_2 | 0, t_1 \rangle_J, \qquad (3.11)$$

where one again retains only the explicit dependence of $W_{2,1}$ of g. The solution of (3.11) is elementary,

$$\langle 0, t_2 | 0, t_1 \rangle_J = \exp \left\{ -ig \int_{t_1}^{t_2} d^4x \left[\left(\frac{1}{i}\right) \delta/\delta J(x) \right]^n \right\} \langle 0, t_2 | 0, t_1 \rangle_J^0$$

or

$$\langle 0, t_1|0, t_1\rangle_J = \exp\left[i\int_{t_1}^{t_2} d^4x\mathcal{L}'\left\{\left(\frac{1}{i}\right)\delta/\delta J(x)\right\}\right]\langle 0, t_2|0, t_1\rangle_J^0. \quad (3.12)$$

This second form of the vacuum amplitude, (3.12), can be generalized immediately to all local, canonical, interacting field theories; and it can be recast into the form of a functional integral [Feynman and Hibbs (1965)], which is frequently convenient both for computation and for the formulation itself when constraints are present.

Before touching upon the latter topic, it is convenient to use (2.42) to rewrite (3.12), in the form

$$\langle 0, \infty|0, -\infty\rangle_J = e^{\frac{i}{2}\int J\Delta_c J} \cdot e^{\mathcal{D}} \cdot e^{i\int d^4x\mathcal{L}'[A]}, \quad (3.13)$$

where, now, $\mathcal{D} = -\frac{i}{2}\int \frac{\delta}{\delta A}\Delta_c\frac{\delta}{\delta A}$ and $A = \int \Delta_c J$.

A perturbation expansion of all the n-point functions of the theory can be obtained by expanding either (3.12) or (3.13) in powers of the coupling, g. Renormalization must always be performed, relating measurable quantities to the input parameters of the theory, m and g, which - in renormalizable theories - serves to remove systematically all the infinities found in the perturbative approximations. For application to SC systems with large couplings, where perturbative techniques are questionable, many attempts have been made to find other, non-perturbative approaches[3]. This is still an open question in the sense that the strong-coupling version of QED, as described in Part 2, appears to be quite different from that of the QCD of Part 3. The last words have not yet been written on this subject.

It is useful to rewrite (3.12) in terms of a FFT, based on (2.28),

$$\langle 0, \infty|0, -\infty\rangle_J^0 = c'\int d[\phi]\exp\left[-\frac{i}{2}\int\phi\Delta_c^{-1}\phi + i\int J\phi\right],$$

with $c'^{-1} = \int d[\phi]\exp\left[-\frac{i}{2}\int\phi\Delta_c^{-1}\phi\right]$. Then, operation by $\exp\left[i\int\mathcal{L}'\left[\frac{1}{i}\frac{\delta}{\delta J}\right]\right]$ upon $\exp[i\int J\phi]$ produces $\exp[i\int\mathcal{L}'[\phi] + i\int J\phi]$ so that

$$\langle 0, +\infty|0, -\infty\rangle_J = c'\int d[\phi]\exp\left[-\frac{i}{2}\int\phi\Delta_c^{-1}\phi + i\int\mathcal{L}'[\phi] + i\int J\phi\right]. \quad (3.14)$$

Note that an integration-by-parts transforms $\int\phi\Delta_c^{-1}\phi$ into $\int[m^2\phi^2 + (\partial_\mu\phi)^2]$; and hence that (3.14) is equivalent to

[3] For example, [Bender, et al. (1979)]. The first of a continuing series of papers is, e.g., [Bender, et al. (1988)].

$$\langle 0, +\infty | 0, -\infty \rangle_J = c' \int d[\phi] \exp \left[i \int \mathcal{L}[\phi] + i \int J\phi \right], \tag{3.15}$$

where $\mathcal{L} = \mathcal{L}_0 + \mathcal{L}'$ denotes the full Lagrangian density.

Another way of constructing (3.15) is to follow a method of Symanzik [Symanzik (1954)] and, independently, of Fradkin [Fradkin (1954)]. One starts with the Euler field equation, converts it and the ETCRs into a differential equation for functional derivatives of $Z[J]$. Symanzik and Fradkin then solved that equation by first finding the solution for $g = 0$, and then making an ansatz for the complete solution which finally appears in Schwinger's form. However, one can proceed in a different way, by writing $Z[J]$ as a FFT with functional Fourier transform $T[f]$. Substitution into the transform, which can then be algebraically obtained, leading directly to (3.15).

This method is somewhat interesting[4] when one is in "non-canonical" situations, where the generalized momenta depend on the fields; then, it is useful to write a double FFT, for the source dependence of J of the fields (conjugate to ϕ) and for an independent source, say K of the momenta (conjugate to χ). One substitutes the double FFT into the functional differential equations, solves the resulting equations for the transform $T[\phi, \chi]$ and performs the necessary functional integration over χ. Then setting $K = 0$, one has automatically the result in the form of a single FFT giving the J-dependence in terms of a FI over ϕ.

3.2 A Gauge Digression: SU(N)-QCD

One very important representation of a GF is that arising when gauge invariance is a non-trivial consideration, in particular for QCD, where the invariance of physical quantities under gauge transformations as written in (5.25) is to be preserved. That is, one requires the FI over fluctuations A_μ^a when one knows beforehand that not all such fields are independent. With (5.25) it is always possible to transform to a new set of fields $A_\mu'^a$ which satisfy a prescribed condition, or set of conditions, that act to constrain or remove dependent field components only. In this Section a simple construction will be given[5] of the FI in this non-Abelian situation, from the

[4]See, e.g., Chapter 4 of [Fried (1972)], where a chiral-invariant pion model is discussed.

[5]The gauge invariance discussed here is treated in much more detail by ['t Hooft and Veltman (1973)]; and in [Veltman (1975)]. The author is indebted to A. Bogojevic and to M. Blagojevic for informative discussions on this subject.

point of view of maintaining gauge invariance; lengthier discussions, including constraints and the conventional Faddeev-Popov construction, may be found in a variety of texts[6].

Let us take the form of (3.15) as the starting point in the construction of a GF for QCD, with the replacements $\phi \to A^a_\mu$, $J \to J^a_\mu$. The Lagrangian $\mathcal{L}[\phi]$ should be replaced by the QCD $\mathcal{L}[A] = -(1/4)(F^a_{\mu\nu})^2$, with $F^a_{\mu\nu} = \partial_\mu A^a_\nu - \partial_\nu A^a_\mu + g f_{abc} A^b_\mu A^c_\nu$, as in (5.24); and $\mathcal{L}$ is, by inspection, invariant under the local gauge transformations (5.25). In the absence of sources, $J^a_\mu = 0$, the amplitude $\langle 0,+\infty|0,-\infty\rangle_{J=0}$ represents the probability amplitude that a vacuum at $t = -\infty$ remains one at $t = +\infty$. Its absolute square is certainly a physical quantity, and as such should be independent of gauge; and in order to assure this, we shall assume that the vacuum amplitude must also be gauge-invariant.

One therefore begins by writing this amplitude as

$$\prod_{a,\mu} \int d[A^a_\mu] \exp\left[i \int \mathcal{L}[A] \right], \tag{3.16}$$

but immediately realizes that this is not yet appropriate, for a choice of gauge can always be made so that certain components $B^a_\mu(z)$ satisfy equations of motion, while the other components, say $C^a_\mu(z)$, satisfy equations of constraint; these equations of constraint may be differential or integral or algebraic. It is only over the independent B variables that one should integrate, and not over the dependent C, for - if the Cs can be eliminated in favor of the Bs - the remainder of the integrand will be independent of the Cs, so that integration over them will just produce an infinite factor. Differentiation between independent and dependent components is, typically, expressed by a gauge condition $\mathcal{F}^a[A] = 0$, where $\mathcal{F}^a[A]$ denotes some operation[7] on all of the A components which can be achieved by gauge choice of form (5.25). To insure that such a condition is maintained while calculating fluctuations, one can insert a product of delta-functionals, $\delta[\mathcal{F}[A]] = \prod_a \delta\mathcal{F}^a[A]$, under the integrand of (3.16). However, this cannot

[6]Treatments dealing with constraints, and extensive references to the literature on this subject, may be found in [Itzykson and Zuber (1980)], and in [Faddeev (1980)]. Perhaps the original reference is [Dirac (1958)].

[7]Simple examples are: the axial gauge, $\eta_\mu A^a_\mu(x) = 0$, where η_μ is a constant 4-vector; the coordinate gauge (discussed in Chapter 13), $x_\mu A^a_\mu(x) = 0$; and the Landau gauge, $\partial_\mu A^a_\mu(x) = 0$. One of the most satisfying, if non-covariant, of all gauges has been discussed by ['t Hooft (1979)], in which all gauge freedom is restricted modulo a global phase factor.

be the entire story, for one still has the requirement of gauge invariance: if a change to another gauge is made, $\mathcal{F}^a[A] \to \mathcal{H}^a[A]$, there will in general be a change in the value of the amplitude so defined.

What is needed is some other functional of A, say $M[A]$, inserted under the integral, which under a gauge transformation changes in just such a way that the product of $\delta[\mathcal{F}[A]]M[A]$ is unchanged. That is, if $M_{\mathcal{F}}[A]$ is associated with the gauge defined by $\mathcal{F}^a[A] = 0$, and if $M_{\mathcal{H}}[A]$ is associated with the gauge defined by $\mathcal{H}^a[A] = 0$, then one should require that

$$\delta[\mathcal{F}[A]]M_{\mathcal{F}}[A] = \delta[\mathcal{H}[A]]M_{\mathcal{H}}[A]. \tag{3.17}$$

Suppose the gauge conditions are related in a very general way by $\mathcal{H}^a[A] = Q_{ab}[A]\mathcal{F}^b[A]$. With the aid of the formula

$$\delta[\mathcal{H}[A]] = [\det Q]^{-1}\delta[\mathcal{F}[A]],$$

equating coefficients of $\delta[F[A]]$ in (3.17) yields

$$M_{\mathcal{H}}[A] = \det(Q)M_{\mathcal{F}}[A] = \det|\delta\mathcal{H}/\delta\mathcal{F}|M_{\mathcal{F}}[A]. \tag{3.18}$$

In order to satisfy (3.18), and hence gauge invariance of the amplitude, it is sufficient that $M_{\mathcal{F}}[A] \sim \det|\delta\mathcal{F}/\delta z|$, where z_a is some parameter or function or functional carrying the color index $\underline{a}$. In the conventional Faddeev-Popov construction, one chooses $z_a(x)$ as the gauge function $\omega_a(x)$ of the infinitesimal transformation $V = \exp[i\lambda^a\omega_a] \to 1 + i\lambda^a\omega_a$, so that $\delta\mathcal{F}/\delta\omega$ is evaluated in the $\omega \to 0$ limit.

We now sketch the argument of 't Hooft and Veltman ['t Hooft and Veltman (1973)] which shows how the entire FI may be reduced to one over a set of fields supposed to be the unique solution to the particular gauge condition used. Denote the latter by the variables $\bar{A}^a_\mu(x)$, while from any such $\bar{A}$ an infinite class of gauge-equivalent fields may be constructed by a transformation of the type (5.25), with gauge function $\omega_a(x)$. One imagines a "gauge space" in which such ω-transformations define "gauge orbits" orthogonal to the directions defined by variation of the $\bar{A}$. The complete space of fluctuations can then be divided into $\bar{A}$ and ω variations: $d[A] = d[\bar{A}]d[\omega]$.

If the $\bar{A}$ are solutions to the equations $\mathcal{F}[A] = 0$, then gauge transformations away from the $\bar{A}$ are effected by the ω; and hence the presence of the factor $\delta[\mathcal{F}[A]]$ multiplying $\delta\mathcal{F}/\delta\omega$ means that the latter functional may be evaluated at $\omega = 0$. One then has

$$\int d[\bar{A}] \int d[\omega] \delta[\mathcal{F}[A]] \det |\frac{\delta \mathcal{F}}{\delta \omega}|$$
$$= \int d[\bar{A}] \int d[\mathcal{F}] \delta[\mathcal{F}] = \int d[\bar{A}], \quad (3.19)$$

where the $d[\mathcal{F}]$ integration in (3.19) is over all the variables lying on gauge orbits orthogonal to variations of the $\bar{A}$. In this way the construction has been shown to be both gauge invariant, and to correspond to functional integration only over the unique $\bar{A}$ variables defined in a particular gauge.

It was pointed out by Gribov [Gribov (1978)] that non-perturbative solutions of the gauge condition $\mathcal{F}[A] = 0$ may not be unique, for certain gauges, leading to the possibility that different solutions $\bar{A}$ may exist in different regions of function space. The FI over $\int d[\bar{A}]$ would then have to be "patched" together, using those appropriate to the different, non-perturbative regions[8]. Such difficulties usually disappear, along with the necessity of retaining an explicit $M[A]$, when one uses an axial or coordinate gauge; in these gauges, $M[A]$ is independent of A, and can be absorbed into the normalization of the FI. In the non-perturbative formulation of QCD described in Part 3, a simple rearrangement of the Schwinger solution is used to produce a manifestly gauge-invariant GF [Fried, et al. (2010)].

3.3 Coupled Fermion and Boson Fields

Let us return to the Abelian world of QED. Although this subject should really be discussed in terms of constraints [Itzykson and Zuber (1980)]; [Faddeev (1980)]; [Dirac (1958)], as long as charge is conserved that complication is not really necessary for the calculation of any physical, gauge-invariant quantity. What we shall do instead is to take the simplest path, justified at the end of this Section, replacing the scalar A of Section A by the vector A_μ, pretending that each component A_μ may be independently varied – as one does in the Feynman gauge, where the causal photon propagator is obtained from that of the scalar problem by appending to it a factor of $\delta_{\mu\nu}$ and taking the limit of zero boson mass. Thus, in the absence of charged fermion sources, and with the end-point times taken into the remote past and future, the generalization of (3.10) becomes

[8] [Mitter (1973)]. See also the stochastic quantization described by [Zwanziger (1985)].

$$Z^{(0)}[J] = \langle 0, \infty | 0, -\infty \rangle^0_{J_\mu} = \exp\left[\frac{i}{2}\int J_\mu D_{c,\mu\nu} J_\nu\right], \tag{3.20}$$

with $D_{c,\mu\nu}(x-y) = \delta_{\mu\nu}\Delta_c(x-y,m)|_{m\to 0}$, the free-photon propagator in the Feynman gauge. There is a class of relativistic gauges into which one can pass, for ease of calculation of certain physical quantities[9]; perhaps the most useful for closed fermion-loop computations is the Landau gauge, which can be defined[10] by the requirement $\partial_\mu D_{c,\mu\nu} = \partial_\nu D_{c,\mu\nu} = 0$, i.e., $D_{c,\mu\nu} = (\delta_{\mu\nu} - \partial_\mu\partial_\nu/\partial^2)\Delta_c(x-y,o)$.

The GF for free fermions may be written in a manner analogous to (3.20), with the aid of anticommuting, c-number (Grassmann) sources η_α, $\bar{\eta}_\beta$ as

$$Z^{(0)}[\bar{\eta},\eta] = \langle 0, \infty | 0, -\infty \rangle^0_{\nu\bar{\nu}} = \exp\left[i\int \bar{\eta} S_c \eta\right]. \tag{3.21}$$

An analysis similar to that leading to (3.8) shows that

$$Z^{(0)}[\bar{\eta},\eta] = \langle 0| \left(\exp\left[i\int[\bar{\eta}\psi + \bar{\psi}\eta]\right]\right)_+ |0\rangle, \tag{3.22}$$

with the definition of ordering for fermion fields as discussed after (1.22). Here, the η_α, $\bar{\eta}_\beta$ are to anticommute with themselves and with all free-fermion fields $\psi^{(0)}_\alpha$, $\bar{\psi}^{(0)}_\beta$, which properties lead to the correct free-fermion propagator of (1.19). [Note that a factor of $\psi^{(0)}_\alpha$ is achieved by operation upon (3.22) with $(\frac{1}{i})\delta/\delta\bar{\eta}_\alpha$; but that a $\bar{\psi}^{(0)}_\beta$ requires operation with $i\delta/\delta\eta_\beta$.

The Lagrangian of QED may be written as

$$\mathcal{L} = \mathcal{L}_0 + \mathcal{L}', \mathcal{L}_0 = \mathcal{L}_0(A) + \mathcal{L}_0[\bar{\psi},\psi],$$

with $\mathcal{L}' = ig\bar{\psi}\gamma\cdot A\psi$. Following the Action Principle analysis which lead to (3.12), (3.21), and (3.22), one obtains

$$\langle 0,\infty|0,-\infty\rangle_{J,\eta,\bar{\eta}} = \exp\left[i\int d^4x\left(-\frac{1}{i}\frac{\delta}{\delta\eta}\right)\gamma\cdot\left(g\frac{\delta}{\delta J}\right)\left(\frac{1}{i}\frac{\delta}{\delta\bar{\eta}}\right)\right] \cdot\exp\left[i\int \bar{\eta}S_c\eta + \frac{i}{2}\int JD_cJ\right], \tag{3.23}$$

[9]The most familiar gauges are (for UV questions) the Landau gauge, described in the text; and (for IR-sensitive calculations) the Yennie gauge, where $D_{c,\mu\nu} = (\delta_{\mu\nu} + 2\partial_\mu\partial_\nu/\partial^2)D_c$. Physical quantities are gauge-invariant; but different computations may be simplified in different gauges.

[10]The change of the QED n-point functions, with changes in the gauge structure of the free-photon propagator, was described by [Zumino (1960)].

with

$$Z\{J,\bar{\eta},\eta\} = \langle 0| \left(\exp\left\{ i \int [\bar{\psi}\eta + \bar{\eta}\psi + JA] \right\} \right)_+ |0\rangle,$$

whose vacuum states are frequently chosen to be the "IN" states discussed in Chapter 4. In this case of interaction, the GF and the vacuum-to-vacuum amplitude differ by a normalization constant, which from (3.23) is just the vacuum amplitude in the limit of zero sources: $\langle 0,+|0,-\rangle = \langle 0,+\infty|0,-\infty\rangle|_{J=\eta=\bar{\eta}=0}$,

$$Z\{J,\bar{\eta},\eta\} = \langle 0,+|0,-\rangle^{-1} \cdot \exp\left[-i \int \frac{\delta}{\delta\eta} \left(-g\gamma \cdot \frac{\delta}{\delta J} \right) \frac{\delta}{\delta\bar{\eta}} \right] \cdot Z^{(0)}\{J,\bar{\eta},\eta\}, \quad (3.24)$$

where $Z^{(0)}\{j,\bar{\eta},\eta\}$ denotes the product of (3.20) and (3.21).

This formal, functional solution for the GF, and hence for all the coupled n-point functions, may be recast into a form both instructive and useful, by applying the fermion Gaussian combinatoric of (2.16). Here, $A_{\alpha\beta}(x,y) = -g\delta(x-y)\delta_\mu^{\alpha\beta}\delta/\delta J_\mu(x)$ while $B_{\alpha\beta}(x,y) = S_c^{\alpha\beta}(x-y)$; and the result is

$$Z\{J,\bar{\eta},\eta\} = \langle 0+|0-\rangle^{-1} \cdot e^{i\int \bar{\eta} G_c[\frac{1}{i}\frac{\delta}{\delta J}]\eta} \,.\, e^{L[\frac{1}{i}\frac{\delta}{\delta J}]} \,.\, e^{\frac{i}{2}\int J D_c J}, \quad (3.25)$$

where

$$G_c[A] = S_c[1 - ig\gamma \cdot AS_c]^{-1},$$

and

$$L[A] = Tr \ln[1 - ig\gamma \cdot AS_c].$$

Finally, we may apply the reciprocity relation of (2.8) to rewrite this in a more compelling way,

$$Z\{J,\bar{\eta},\eta\} = e^{\frac{i}{2}\int J D_c J} \,.\, e^{\mathcal{D}} \,.\, e^{i\int \bar{\eta} G_c[A]\eta} \,.\, e^{L[A]} / \langle 0+|0-\rangle, \quad (3.26)$$

where $\mathcal{D} = -\frac{i}{2}\int \frac{\delta}{\delta A_\mu} D_{c,\mu\nu} \frac{\delta}{\delta A_\nu}$, $A_\mu(x) = \int D_{c,\mu\nu}(x-y)J_\nu(y)$ and where all indices have been elsewhere suppressed. Note that from (3.24), the vacuum-to-vacuum amplitude is itself given by

$$\langle 0+|0-\rangle = e^{\mathcal{D}} \cdot e^{L[A]}|_{A\to 0}. \tag{3.27}$$

This form of the GF is instructive, for it clearly points out the difference between the "first" quantization of Potential Theory, and the "second" quantization of QFT. In the former, one is typically interested in the solution — in a form as close to explicit as possible — for $G_c(x,y|A) = \langle x|G_c[A]|y\rangle$, the fermion propagator in a given, explicit background field. The n-point functions of second quantization, however, are given by linkages (or FIs over gaussian-weighted fluctuations) over products of such $G_c[A]$, with all possible numbers of closed-fermion-loop $L[A]$ insertions.

The form of (3.27) is also useful, for while it may be converted to a FI, as in Chapter 2, it is somewhat simpler to express such operations in terms of linkages; and this is what shall be done almost everywhere in this book.

One can in this way clearly appreciate the complexities of quantum field theory: one is first required to find appropriate forms for the first-quantization functionals $G_c[A]$ and $L[A]$; and then one must perform the necessary linkages, or FIs, over multiple products of these functionals. Gauge invariance of this formulation of QED is restored when calculating physical S-Matrix elements for any process, as the gauge ambiguities obtained from the GF are removed by the process of "mass-shell amputation". This is described in detail in Chapter 4 of *Functional Methods and Models in Quantum Field Theory* [Fried (1972)], and in Part 2, below.

3.4 Fields at the Same Point

An ambiguity in all previous discussions is reflected in the lack of precision given to the definition of operator fields, or Green's function arguments, at the same space-time point. Such quantities are usually quite singular, while the methods of introducing a reasonable degree of finiteness are not always the same for all interactions.

The simplest level at which this problem arises concerns the definition of the free-field boson GF, (3.10). The elementary problem now considered is the use of the Action Principle to obtain the GF for $m \neq 0$ is it is already known for $m = 0$. For arbitrary m, one has from (1.4), and in the manner of (3.3),

$$\frac{\delta}{\delta m^2}\langle 0,+\infty|0-\infty\rangle_J = -\frac{i}{2}\int d^4x\langle 0,+\infty|A^2(\alpha)|0,-\infty\rangle_J$$
$$= \frac{i}{2}\int d^4x\left[\frac{\delta}{\delta J(x)}\right]^2\langle 0,+\infty|0,-\infty\rangle_J, \quad (3.28)$$

with solution

$$\langle 0,\infty|0,-\infty\rangle_J = \exp\left[-\frac{i}{2}\int\frac{\delta}{\delta J}[-m^2]\frac{\delta}{\delta J}\right]\cdot\langle 0,+\infty|0,-\infty\rangle_J^{m=0} \quad (3.29)$$

where $\langle x|m^2|y\rangle = m^2\delta(x-y)$. From (3.20), written for a scalar massless field, one has

$$\langle 0,+\infty|0,-\infty\rangle_J^{m=0} = \exp\left[\frac{i}{2}\int JD_cJ\right], \quad (3.30)$$

where $D_c(x) = \Delta_c(x;m)|_{m=0}$, while (2.13) permits the evaluation of the RHS of (3.29),

$$\exp\left[\frac{i}{2}\int JD_c(1+[m^2]D_c)^{-1}J\right]\cdot\exp\left[\frac{1}{2}Tr\ln(1+[m^2]D_c)^{-1}\right]. \quad (3.31)$$

The J-dependent term of (3.31) is exactly that of (3.30), since

$$\langle p|D_c(1+m^2D_c)^{-1}|p'\rangle = \langle p|p'\rangle\vec{D}_c(p)[1+m^2\vec{D}_c(p)]^{-1}$$
$$= \delta(p-p')[m^2+p^2]^{-1},$$

which was the starting point of the calculation. The remaining term of (3.31) is just an (infinite) phase factor[11], but its existence rather than its value is the point requiring an explanation.

A clue to the origin of this discrepancy lies in the interpretation of the phase as a vacuum-to-vacuum effect, for this change of mass shifts the energy level of the vacuum by an infinite amount. A similar effect occurs in conventional field theory, where one calculates a free-field Hamiltonian directly from a free-field Lagrangian, and discards the "infinite zero-point energy" as a meaningless constant, since only energy differences can be physical. That part of the discarded constant independent of spatial momenta

[11]This may be seen by supplying sufficient regulation of $\tilde{D}_c(p)$ so that the integral defining $\ln[N]$ may be computed, after which the regularization is removed. For Example, one may replace $[p^2]^{-1}$ by $2\int_0^{\Lambda^2} LdL(L+p^2)^{-3}$, and subsequently take the limit of very large Λ^2.

has returned to generate the unwanted phase of (3.31), as the self-energy of the vacuum tries to adjust to its new infinite value when m increases from zero.

The original Lagrangian may be rewritten so that it does not generate this zero-point effect by subtracting away the vacuum expectation value of $\mathcal{L}$, replacing $\mathcal{L}_0(A)$ by $\mathcal{L}_0(A) - \langle 0|\mathcal{L}_0(A)|0\rangle$. Translational invariance requires that $\langle 0|A^n(x)|0\rangle$ be independent of x; and hence this subtraction removes a constant from the term $-(m^2/2)A^2(x)$ of the free Lagrangian, replacing the latter by $-(m^2/2)[A^2(x) - \langle 0|A^2|0\rangle]$. For free fields, as seen in the next Chapter, this can be written as $-(m^2/2)\langle 0| : A^2(x) : |0\rangle$, the Wick- or normal-ordered product, in which creation operators stand to the left and destruction operators to the right. As in (3.5), $A^2(x)$ is always to be understood as the symmetric limit of a time-ordered product, and hence $\langle 0|A^2(x)|0\rangle$ may be denoted by $-i\Delta_c(0)$, a divergent constant. Application of the Action Principle now yields

$$\frac{\delta}{\delta m^2}\langle\, 0\, , +\infty|0, -\infty\rangle_J$$
$$= -\frac{i}{2}\int d^4x \langle 0, +\infty|[A^2(x) - \langle 0|A^2(x)|0\rangle]|0, -\infty\rangle$$

in place of (3.28), with the previous solution (3.29) now multiplied by an extra factor

$$\exp\left[-\frac{i}{2}d\mu^2 \int_0^{m^2} \int d^4x \langle 0|A^2|0\rangle_{\mu^2}\right] = \exp\left[\frac{VT}{2}\int_0^{m^2} d\mu^2 \Delta_c(0;\mu^2)\right],$$

which exactly cancels the phase factor of (3.31). In this way, while providing the definition of what is meant by a pair of free field operators at the same point, the normal-ordering prescription restores the validity of the Action Principle analysis.

In a theory with interactions of form $\mathcal{L}' = -gA^n$, one must supply a corresponding definition of n operators acting at the same point. One natural method is with the aid of a Wick-ordering operator functional [Symanzik (1961)],

$$W[J] =: \exp[i\int JA] :\equiv \left(\exp[i\int JA]\right)_+ \langle 0|\left(\exp[i\int JA]\right)_+|0\rangle^{-1} \tag{3.32}$$

where

$$: A(x_1) \ldots A(x_n) :\equiv \frac{1}{i}\frac{\delta}{\delta J(x_1)} \cdots \frac{1}{i}\frac{\delta}{\delta J(x_n)} \cdot W[J]|_{J=0}. \tag{3.33}$$

As an example, these equations applied to this interaction replace $\mathcal{L}'$ by

$$: \mathcal{L}' := -g : A^4(\alpha) := -g\left\{A^4 - \langle A^4\rangle + 6\langle A^2\rangle(A^2 - \langle A^2\rangle)\right\} \tag{3.34}$$

under the assumption that $\langle 0|A|0\rangle$ vanishes for interacting fields as it does for free fields. The use of $: \mathcal{L} :$ rather than $\mathcal{L}$ is therefore equivalent to using $\mathcal{L}$ with the infinite constants

$$-g\{\langle A^4\rangle + 3!\langle A^2\rangle^2 + 12m^2\langle A^2\rangle\}$$

removed, together with a shift of the unrenormalized mass m by the amount: $(3!)g\langle 0|A^2|0\rangle$. Quite generally, the use of $: \mathcal{L} :$ for any polynomial interaction corresponds to the removal of the vacuum self-energy infinity together with the redefinition of the unrenormalized couplings. It should be noted that there exists alternate "point splitting" prescription for the removal of such infinities [Levy (1964)];[Zimmerman (1967)];[Zimmerman (1968)].

For fermion fields, and in particular in QED, the situation is more complicated but more amenable to solution, since there exist certain principles which may be invoked to define a proper interaction. Normal ordering of both $\mathcal{L}_0[A_\mu]$ and $\mathcal{L}_0[\bar{\psi}, \psi]$ may be adopted, but this prescription leaves untouched the fermion current operator $j_\mu = ig\bar{\psi}\gamma_\mu\psi$, with two fields at the same point, which when multiplied by the photon field A_μ at that space-time point, is written for the interaction term, $\mathcal{L}' = j_\mu A_\mu$. Not only should the vacuum energy be zero, but in addition its current density must vanish in the limit of zero photon field,

$$\langle j_\mu(x)\rangle_{A_\nu=0} = 0. \tag{3.35}$$

Further, for any value of A_μ, the current so induced in the vacuum by an external field must be conserved,

$$\partial_\mu\langle j_\mu(x)\rangle_{A_\nu} = 0, \tag{3.36}$$

in order to guarantee that the total charge of the vacuum remains zero. These requirements are stated for the VEV of the current operator, and consequently it is for this quantity — rather than the operator directly — that the analysis is performed.

The simplest definition of the current operator would be given by

$$j_\mu(x) = i \lim_{x-x'\to 0} \sum_{\alpha\beta} \gamma_\mu^{\alpha\beta} (\bar{\psi}_\beta(x')\psi_\alpha(x))_+, \tag{3.37}$$

where, to maintain strict causality, the points x and x' should be relatively space-like. In this case, an alternate form of (3.37) is obtained by averaging over both orderings of these factors, to produce

$$\frac{i}{2} \lim_{x-x'\to 0} \sum_{\alpha\beta} \gamma_\mu^{\alpha\beta} [\bar{\psi}_\beta(x'), \psi_\alpha(x)];$$

for free fermions, this is equivalent to the normal-ordered form. Were these fermion fields dependent upon external, *c*-number (background) sources A_μ, upon taking the VEV of (3.37) there would result

$$\langle j_\mu(x)\rangle_{gA} = - \lim_{x-x'\to 0} tr[\gamma_\mu G_c(x, x'|gA)] \tag{3.38}$$

where *tr* means a trace over Dirac indices, and the GF of (3.23) has been used to identify $G_c[A]$ with this time-ordered quantity. Comparing our previous QED forms with (3.38), one sees the connection between $\langle 0|j_\mu|0\rangle$ and the closed-fermion-loop functional $L[A]$,

$$L[A] = i \int_0^g dg' \int d^4x A_\mu(x) \langle j_\mu(x)\rangle_{g'A}. \tag{3.39}$$

An alternate and useful relation can be seen to follow from these definitions,

$$\frac{\delta L}{\delta A_\mu(x)} = ig\langle j_\mu(x)\rangle_{gA}. \tag{3.40}$$

Hence any properties and requirements placed upon the current induced in the vacuum by an external field will have an immediate application to and effect on $L[A]$, and thereafter to all photon processes.

It should be remarked that, formally, there is no difficulty at all: if the fermion field equations corresponding to the interaction $\mathcal{L}'$ are satisfied, one expects (3.36) and (3.35) to be satisfied if only the limit $x - x' \to 0$ is taken in a symmetric way. The difficulty is that $G_c(x, x'|A)$ becomes singular in this limit, and formal manipulations can go awry. The resolution of this problem, originally given by Valatin [Valatin (1954)] will here be accomplished by first changing the definition of $\langle 0|j_\mu|0\rangle_{gA}$ to insure that

$$\partial_\nu^y \frac{\delta}{\delta A_\nu(y)} \langle j_\mu(x)\rangle_{gA} = 0 \tag{3.41}$$

for any coordinate y; and then employing translational invariance to argue that (3.41) is equivalent to (3.36).

Under gauge transformations of the form $A \to A_\mu + \partial_\mu \Lambda$, the Green's function $G_c[A]$ will undergo the transformation

$$G_c(x,y|gA) \to G_c(x,y|g[A+\partial\Lambda]) = G_c(x,y|gA) \cdot \exp[ig(\Lambda(x) - \Lambda(y))] \tag{3.42}$$

as is easily seen by examining the DEs obeyed by $G_c[A]$; this will also be demonstrated using the detailed Schwinger/Fradkin representation for $G_c[A]$ in Chapter 4.4. Hence the combination

$$G_F(x,x'|A) \equiv G_c(x,x'|A) \cdot \exp\left[-ie\int_{x'}^{x} d\xi_\sigma A_\sigma(\xi)\right] \tag{3.43}$$

is explicitly independent of Λ under this gauge transformation, which condition implies that

$$\partial_\nu^y \frac{\delta}{\delta A_\nu(y)} G_F(x,x'|A) = 0.$$

Thus if the $G_c[A]$ in the original definition of (3.38) is replaced by $G_F[A]$,

$$\langle j_\mu|x\rangle_F \equiv - \lim_{x-x'\to 0} tr[\gamma_\mu G_c(X,x')A] \cdot \exp\left[-ie\int_{x'}^{x} d\xi_\sigma A_\sigma(\xi)\right], \tag{3.44}$$

(3.41) will be satisfied by (3.44) for any point y and for any separation $x-x'$. Taken by itself, the exponential factor of (3.44) vanishes as $x - x' \to 0$; but it multiplies $G_c(x,x'|A)$ which is singular in that limit; and their product, while not itself finite, is just sufficient to insure the gauge invariance of the combination.

The nature of the limit in (3.44) should (i) be such that the coordinate difference is always space-like, and (ii) maintains $\langle 0|j_\mu|0\rangle_F$ as a 4-vector while (iii) preserving the translational invariance of all resulting amplitudes which arise from the further expansion of $\langle 0|j_\mu|0\rangle_F$ in (odd) powers of A. In perturbative expansions of QED these properties are verified *a posteriori*, while they are explicitly exhibited in (zero fermion mass) solutions of two-dimensional QED. If properties (ii) and (iii) are fulfilled, then $\langle 0|j_\mu|0\rangle_F$ may be written in terms of its Taylor expansion,

$$\int K^{(2)}_{\mu\nu}(x-y)A_\nu(y) + \int K^{(4)}_{\mu\nu\lambda\sigma}(x-y,y-z,z-w)A_\nu(y)A_\lambda(z)A_\sigma(w)+\dots$$

and (3.41) employed to show that each $K^{(2\ell)}_{\mu\nu\dots}$ satisfies $\partial^x_\mu K^{(2\ell)}_{\mu\nu\dots}=0$; and hence (3.36) is satisfied. An operator redefinition of j_μ which yields (3.44) can easily be invented [Fried (1972)].

Chapter 4

The Generating Functional and the S-Matrix

The probability amplitudes, or S-matrix elements, which describe transitions between states of physical systems can be presented in the form of certain mass-shell operations upon appropriate multi-point Green's functions, (1.21). The latter can be most succinctly obtained from a generating functional (GF) which in turn will permit the so-called reduction formulae of the S-matrix to be exhibited in a clear and compact way.

4.1 The Generating Functional Operator

It is simplest to begin by considering[1] a scalar, hermitian boson field $A(x)$, with the generalization to other types of fields performed at a later stage. A "generating functional operator" labeled by two space-like surfaces σ_a and σ_b is defined as in (3.8) by

$$\overset{a}{\mathcal{T}^b}\{j\} = \left(e^{i\int_b^a jA}\right)_+, \tag{4.1}$$

where

$$\int_b^a jA \equiv \int_{\sigma_b}^{\sigma_a} d^4x j(x)A(x),$$

with $j(x)$ denoting an arbitrary, real, c-number source function. The surfaces $\sigma_{a,b}$ may be thought of as flat time-cuts, for which the integral of (4.14) becomes

$$\int d^3x \int_{t_b}^{t_a} dt j(x)A(x),$$

[1]The discussion of this Section follows that given by [Symanzik (1960)].

with $t_a \geq t_b$, and the volume integral extending over all space. The time-ordering prescription of (4.1) is the same as that of (3.6).

The functional derivative (FD) of (4.1) may be written in the form

$$\frac{\delta}{\delta j(x)} \mathcal{T}^{\overset{a}{b}}\{j\} = i(A(x)e^{i\int_b^a jA})_+, \tag{4.2}$$

or

$$\frac{\delta}{\delta j(x)} \mathcal{T}^{\overset{a}{b}}\{j\} = i\mathcal{T}^{\overset{a}{x}} A(x) \mathcal{T}^{\overset{x}{b}} \tag{4.3}$$

if the point x lies on a space-like surface between σ_a and σ_b (or if $t_a \geq x_0 \geq t_b$), and is zero otherwise. Equation (4.3) is a sometimes useful way of expressing the ordering requirements of (4.2).

Variation of $\mathcal{T}^{\overset{a}{b}}$ due to an infinitesimal change of the surface σ_a, about the point x_a, may be written in the form

$$\delta \mathcal{T}^{\overset{a}{b}} = i\delta\sigma(x_a) \cdot j(x_a)(A(x_a)e^{i\int_b^a jA})_+$$

or

$$\frac{\delta \mathcal{T}^{\overset{a}{b}}}{\delta\sigma(x_a)} = ij(x_a)A(x_a)\mathcal{T}^{\overset{a}{b}} \tag{4.4}$$

where the second line of (4.4) follows from (4.3) and the property $\mathcal{T}^{\overset{c}{c}} = 1$. Repeating these steps for a small variation of σ_a about y_a yields

$$\frac{\delta}{\delta\sigma(y_a)} \frac{\delta}{\delta\sigma(x_a)} \mathcal{T}^{\overset{a}{b}} = i^2 j(x_a)j(y_a)A(x_a)A(y_a) \cdot \mathcal{T}^{\overset{a}{b}}. \tag{4.5}$$

Had these variations been performed first at y_a and then at x_a, the result would have been (4.5) with the labels interchanged; hence, the difference of these procedures gives

$$\left[\frac{\delta}{\delta\sigma(x_a)}, \frac{\delta}{\delta\sigma(y_a)}\right] \mathcal{T}^{\overset{a}{b}} = i^2 j(x_a)j(y_a)[A(x_a), A(y_a)]\mathcal{T}^{\overset{a}{b}}. \tag{4.6}$$

If the points x_a, y_a are on the same space-like surface, as assumed, the LHS of (4.6) must vanish, and this implies that the commutator $[A(x), A(y)]$ must vanish space-like. Thus, micro-causality is a necessary condition for the integrability of $\mathcal{T}^{\overset{a}{b}}$.

The adjoint functional operator $\overset{a}{\mathcal{T}}{}^{+}_{b}$ may be written as

$$\overset{a}{\mathcal{T}}{}^{+}_{b}\{j\} = (e^{-i\int_b^a jA})_-, \tag{4.7}$$

where $()_-$ denotes a reversed time-ordered bracket. With (4.7), it is simple to demonstrate that $\overset{a}{\mathcal{T}}_{b}$ is a unitary operator, which property is essential if the S-matrix is to be expressed in terms of $\overset{+\infty}{\mathcal{T}}_{-\infty}$. Because one has, immediately,

$$\frac{\delta}{\delta\sigma(x_a)}(\overset{a}{\mathcal{T}}{}^{+}_{b}\overset{a}{\mathcal{T}}_{b}) = 0, \tag{4.8}$$

the combination $\overset{a}{\mathcal{T}}{}^{+}_{b}\overset{a}{\mathcal{T}}_{b}$ is independent of σ_a; hence it can be evaluated at any space-like surface, in particular $\sigma_a = \sigma_b$, which provides the relation $\overset{a}{\mathcal{T}}{}^{+}_{b}\overset{a}{\mathcal{T}}_{b} = 1$. By considering variations of a point on the surface σ_b, one may prove in the same manner that $\overset{a}{\mathcal{T}}_{b}\overset{a}{\mathcal{T}}{}^{+}_{b} = 1$, which completes the demonstration of unitarity. In what follows we shall be concerned with that operator obtained by letting σ_a and σ_b recede to $+\infty$ and $-\infty$, respectively, and this quantity will be written simply as $\mathcal{T}$.

The usefulness of $\mathcal{T}\{j\}$ is that it permits one to construct all products of time-ordered operators, by performing a suitable number of functional differentiations with respect to the source $j(z)$, and then setting the source equal to zero. Only the vacuum expectation values of such products are actually needed, and for this one defines the generating functional $Z\{j\} = \langle\mathcal{T}\{j\}\rangle$. More generally, one should include other fields in Z in order to be able to construct n-point functions of the form of (3.22). Fermions may be introduced with the aid of anti-commuting c-number sources $\eta_\alpha(x), \bar{\eta}_\beta(y)$, which anticommute with themselves and all fermion fields, $\psi, \bar{\psi}$. In contrast to the natural property valid for boson sources,

$$\left[\frac{\delta}{\delta j(x)}, j(y)\right] = \delta(x-y), \tag{4.9}$$

the fermion sources are to satisfy [Schwinger (1951)]; [Schwinger (1954)]; [Schwinger (1956)]:

$$\left\{\frac{\delta}{\delta\eta_\alpha(x)}, \eta_\beta(y)\right\} = \left\{\frac{\delta}{\delta\bar{\eta}_\alpha(x)}, \bar{\eta}_\beta(y)\right\} = \delta_{\alpha\beta}\delta(x-y), \tag{4.10}$$

with all other combinations anticommuting,

$$\left\{\frac{\delta}{\delta\eta},\frac{\delta}{\delta\bar{\eta}}\right\}=\left\{\frac{\delta}{\delta\eta},\frac{\delta}{\delta\eta}\right\}=\left\{\frac{\delta}{\delta\eta},\psi\right\}=\left\{\frac{\delta}{\delta\eta},\bar{\psi}\right\}=0, \tag{4.11}$$

as in Chapter 2. Then, functional differentiation of the generating functional operator

$$\mathcal{T}\{j,\eta,\bar{\eta}\}=(e^{i\int[jA+\bar{\psi}\eta+\bar{\eta}\psi]})_{+} \tag{4.12}$$

will produce the products(1.21) with the correct change of sign under fermion permutation, as in (3.22). The complete generating functional for scalar bosons and fermions is then taken as $Z\{j,\eta,\bar{\eta}\}=\langle\mathcal{T}\{j,\eta,\bar{\eta}\}\rangle$.

4.2 Asymptotic Conditions

As yet nothing has been said about the nature of the interaction coupling the fields A and $\psi,\bar{\psi}$ since it is only necessary to assume that particle-like solutions exist to the coupled Green's function equations, in order to define the S-matrix. It shall be here assumed that the Fourier transforms of both fermion and boson propagators have, as a function of their relevant invariant variable, a single pole occurring at the appropriate physical masses, with residues Z_2 and Z_3, respectively. An equivalent physical statement is that, in the remote past and future, one is dealing with particles sufficiently separated to be essentially noninteracting and this suggests that the fields of interest can, in such asymptotic regions, be replaced by free fields describing free particles carrying their observed, physical masses. These are the IN- and OUT-fields of Yang and Feldman [Yang (1950)], which are conventionally related to the asymptotic, unrenormalized field operators by the postulates of the "weak asymptotic condition",

$$\begin{aligned}\lim_{x_0\to{-\infty \atop +\infty}}\left[\langle a|A(x)|b\rangle-Z_3^{1/2}\langle a|A_{\text{IN} \atop \text{OUT}}(x)|b\rangle\right]=0,\\ \lim_{x_0\to{-\infty \atop +\infty}}\left[\langle a|\psi(x)|b\rangle-Z_2^{1/2}\langle a|\psi_{\text{IN} \atop \text{OUT}}(x)|b\rangle\right]=0,\end{aligned} \tag{4.13}$$

where $|a\rangle,|b\rangle$ are arbitrary, time-independent states of the system under consideration, specified by the sets of quantum numbers a,b. We will follow previous practice and use a complete set of IN-particle states, $|a\rangle=|a\rangle_{\text{IN}}$;

in particular, the vacuum state used to define Z is constructed with the IN-vacuum state, $\rangle = |0\rangle_{\text{IN}}$.

Such IN- and OUT-states are constructed in terms of Fourier transforms of products of $A_{\text{IN,OUT}}$ operators acting upon the vacuum states $|0\rangle_{\text{IN,OUT}}$, with the latter defined by the limiting operations of time-dependent (Schrödinger) states,

$$|0\rangle_{\substack{\text{IN}\\ \text{OUT}}} = \lim_{t \to \substack{-\infty \\ +\infty}} |0, t\rangle, \tag{4.14}$$

with all external c-number sources set equal to zero. Equation (4.13) must be written as a relation between matrix elements, in order to prevent the contradiction that would appear if a "strong" asymptotic condition were to be assumed; by the latter relation one means an operator statement such as

$$\lim_{t \to -\infty} \left[A(x - Z_3^{1/2} A_{IN}(x) \right] = 0,$$

which has, as an immediate consequence, the incorrect relation $Z_3 = 1$, as for free fields. The presence of the $Z_{2,3}$ in (4.13) is a reflection of the self-interation which the particles must undergo, even when they are sufficiently separated so that there is little interaction between them. It is the removal of these factors, along with a shift of mass generated by the same self-interaction, that corresponds to the process of renormalization. Another and more convenient way of writing (4.13) is in the form

$$A(x) = Z_3^{1/2} A_{\substack{\text{IN}\\ \text{OUT}}}(x) + \int d^4y \Delta_{\substack{R\\A}}(x - y) K_y A(y), \tag{4.15}$$

with (4.15) always understood as a relation between matrix elements. Here, $K_x = m^2 - \partial^2$; and m denotes the physical, renormalized mass, with $K_x A_{\text{IN}}(x) = K_x A_{\text{OUT}}(x) = 0$.

4.3 The S-Matrix

Conventionally defined as that operator which transforms IN states and operators into OUT states and operators,

$$A_{\text{OUT}}(x) = S^+ A_{\text{IN}}(x) S, \quad |a\rangle_{\text{OUT}} = S^+ |a\rangle_{\text{IN}}, \tag{4.16}$$

matrix elements of this operator are interpreted as the probability amplitude for a system originally in a state labeled by quantum numbers $\underline{a}$ at $t = -\infty$ to end up in a state $\underline{b}$ at $t = +\infty$,

$$S_{ba} \equiv_{\text{OUT}} \langle b|a\rangle_{\text{IN}} =_{\text{IN}} \langle b|S|a\rangle_{\text{IN}}.$$

In the Schwinger formalism sketched in the previous Chapter, one constructs matrix elements of asymptotic particle states by functional differentiation of $\langle 0, t_1|0, t_2\rangle_{j\neq 0}$ with respect to (renormalized) sources of corresponding asymptotic arguments, subsequently taking the limit of zero source strength; and one finds, in effect, that the limit of $\langle b, t_1|a, t_2\rangle$ as $t_1 \to +\infty, t_2 \to -\infty$ is equivalent to ${}_{\text{OUT}}\langle b|a\rangle_{\text{IN}}$ for arbitrary states $\underline{a}, \underline{b}$.

Because the A_{IN} and the A_{OUT} are to satisfy the same free-field equations of motion, and are to have the same ETCRs, one infers the existence of a unitary transformation between them as in (4.16), an inference which could be proven if there were but a finite number of degrees of freedom in the problem. We shall here assume (4.16) for both boson and fermion IN- and OUT-fields, and for the states constructed from them.

To demonstrate the connection between the S-matrix and the multipoint Green's functions obtained by functional differentiation of $Z\{j\}$, it is simplest to consider first the case of a single scalar boson field, and its corresponding GF operator $\mathcal{T}\{j\}$. From (4.3) there follows

$$\frac{\delta \mathcal{T}}{\delta j(x)} = i\mathcal{T}^{\infty}_{x} A(x)\mathcal{T}^{x}_{-\infty}, \tag{4.17}$$

and from (4.13),

$$\begin{aligned} \lim_{x_0 \to -\infty} \left[\tfrac{\delta \mathcal{T}}{\delta j(x)} - iZ_3^{1/2}\mathcal{T}A_{\text{IN}}(x)\right] &= 0, \\ \lim_{x-0 \to +\infty} \left[\tfrac{\delta \mathcal{T}}{\delta j(x)} - iZ_3^{1/2}A_{\text{OUT}}(x)\mathcal{T}\right] &= 0, \end{aligned} \tag{4.18}$$

again to be understood as a relation between matrix elements. Statements equivalent to (4.18) may be written in the form of (4.15),

$$\begin{aligned} \tfrac{\delta \mathcal{T}}{\delta j(x)} &= iZ_3^{1/2}\mathcal{T}A_{\text{IN}}(x) + \textstyle\int d^4y \Delta_R(x-y)K_y \tfrac{\delta \mathcal{T}}{\delta j(y)}, \\ \tfrac{\delta \mathcal{T}}{\delta j(x)} &= iZ_3^{1/2}A_{\text{OUT}}(x)\mathcal{T} + \textstyle\int d^4y \Delta_A(x-y)K_y \tfrac{\delta \mathcal{T}}{\delta j(y)}, \end{aligned} \tag{4.19}$$

where the difference of this last pair yields the relation

$$Z_3^{1/2}(A_{\rm OUT}(x)\mathcal{T} - \mathcal{T}A_{\rm IN}(x)) = i\int d^4y \cdot \Delta(x-y)K_y\frac{\delta\mathcal{T}}{\delta j(y)}. \qquad (4.20)$$

Multiplying both sides of (4.2) by S, and introducing (4.16), one obtains

$$[A_{IN}(x), S\mathcal{T}] = iZ_3^{-1/2}\int d^4y\Delta(x-y)K_y\frac{\delta}{\delta j(y)}(S\mathcal{T}), \qquad (4.21)$$

which must be satisfied by the combination $S\mathcal{T}\{j\}$.

At this stage it is useful to define the normal- or Wick-ordered exponential,

$$:\exp\left[\int A_{\rm IN}f\right]:=\exp\left[\int A_{\rm IN}^{(-)}f\right]\exp\left[\int A_{\rm IN}^{(+)}f\right],$$

where $\int A_{IN}^{(\pm)}f = \int d^4zA_{IN}^{(\pm)}(z)f(z)$. Since the commutator of $A_{IN}^{(+)}$ with $A_{IN}^{(-)}$ us a c-number, one may easily verify[2] the relation

$$\left[A_{\rm IN}(x), :e^{\int A_{\rm IN}f}:\right] =: e^{\int A_{\rm IN}f}: i\int d^4y\Delta(x-y)f(y), \qquad (4.22)$$

and a comparison with (2.21) then suggests the form[3]

$$S\mathcal{T} =: e^{Z_3^{-1/2}\int A_{\rm IN}\vec{K}\frac{\delta}{\delta j}}: F\{j\}. \qquad (4.23)$$

It should perhaps be emphasized that the Klein-Gordon operator K_y in (4.23) is to act upon the coordinate freed by the functional differentiation operation $\delta/\delta j(y)$, and not (backwards) upon A_{IN} (which would give a zero result).

Because of the property $\langle : \exp[\int A_{IN}f] :\rangle = 1$, it follows from (4.23) that $F\{j\} = \langle S\mathcal{T}\{j\}\rangle$. Further, if there are no external fields present, the OUT-vacuum state, $_{OUT}\langle 0| =_{IN} \langle 0|$, is physically indistinguishable from the IN-vacuum state, which means that the two can differ only by

[2]This follows from the Baker-Campbell-Hausdorff relation $\exp[a+b] = \exp[a]\cdot\exp[b]\cdot\exp\left(-\frac{1}{2}[a,b]\right)$, valid when $[a,b]$ is a c-number (that is, when $[a,b]$ commutes with $\underline{a}$ and $\underline{b}$). The relation can be derived by writing and solving a simple DE for $F(\lambda) = \exp[-\lambda a]\exp[\lambda(a+b)]$. A comparison with the same relation with $\underline{a}$ and $\underline{b}$ interchanged yields $\exp[a]\cdot\exp[b] = \exp[b]\cdot\exp[a]\cdot\exp[a,b]$, and extracting the linear $\underline{a}$-dependence of this produces $[a,e^b] = [a,b]e^b$, which has been used in (4.22).

[3]In this way, Wick's theorem [Wick (1950)] becomes a statement describing the normal-ordering of an exponential functional operator.

a phase, $_{OUT}\langle 0|$ = $\exp(i\phi)$ with ϕ a real (and usually infinite) number. If, however, one is dealing with a situation in which it is convenient to introduce an external field A_{ext}, produced by an external current J_{ext} with Fourier components capable of creating particles (e.g. a radio antenna), then the OUT-vacuum will receive contributions from all the IN-states, and the vacuum expectation value of the S-matrix is not simply a phase factor. In this case it is appropriate to generalize the previous definition of $Z = \langle S\rangle Z\{j, J_{ext}\} = \langle S\mathcal{T}\rangle$, for this turns out to be the quantity directly calculable from the field equations and ETCRs. One then has the general definition

$$S = \langle S\{J_{ext}\}\rangle : e^{Z_3^{-1/2} \int A_{\mathrm{IN}}\vec{K}\frac{\delta}{\delta j}} : Z\{j, J_{ext}\}|_{j=0}, \tag{4.24}$$

where the non-external sources $j(z)$ are to be set equal to zero after all the functional differentiation operations of (2.24) have been performed. With this generalized "reduction formula", one exhibits the S-matrix in terms of Wick-ordered A_{IN}-fields and associated operations upon the GF. When fermions are included the entire analysis goes through in a similar manner, with the result

$$\frac{S}{\langle S\rangle} = : \exp\left[Z_3^{-1/2}\int A_{\mathrm{IN}}\vec{K}\frac{\delta}{\delta j} + Z_2^{-1/2}\int (\bar{\psi}_{\mathrm{IN}}\vec{\mathcal{D}}\frac{\delta}{\delta\bar{\eta}} - \frac{\delta}{\delta\eta}\overleftarrow{\bar{\mathcal{D}}}\psi_{\mathrm{IN}}\right]$$
$$: Z\{j,\eta,\bar{\eta}\}|_0, \tag{4.25}$$

with the artificial sources $j, \eta, \bar{\eta}$ set equal to zero after all the functional operations are performed.

For the special case of free fields, the GF operator $\mathcal{T}^0\{j\}$ may be evaluated in a way that is useful in a variety of applications. In the absence of interactions, $S = 1$ and $Z_3 = 1$; and (4.23) then provides

$$\mathcal{T}^{(0)}\{j\} =: e^{\int A_{IN}K\frac{\delta}{\delta j}} : e^{\frac{i}{2}\int j\Delta_c j},$$

which is easily evaluated to yield

$$(e^{i\int jA_{IN}})_+ =: e^{i\int jA_{IN}} : e^{\frac{i}{2}\int j\Delta_c j}.$$

This should be compared with the corresponding expression of the unordered exponential,

$$e^{i\int jA_{IN}} =: e^{i\int jA_{IN}} : e^{-\frac{1}{4}\int j\Delta_{(1)}j}.$$

4.4 A Bremsstrahlung Example

In calculating S-matrix elements, the $Z_{2,3}$ factors of (4.25) are always absorbed into the definition of the renormalized charge appearing in the (usually perturbative) expansion of the Green's functions which are used to construct the n-point functions of the GF [4] A useful example of the way in which this reduction formula may be employed occurs in the process of multiple bremsstrahlung. The simplest such exercise is to calculate probability amplitude for the emission of n photons, when an electron is scattered by an external field; the resolution of the IR difficulties of this problem, of fundamental importance in QED, has provided much of the structure appropriate to subsequent relativistic eikonal formulations.

One requires the S-matrix element

$$\langle p', s'; k_1, \epsilon_{\mu_1}(k_1); \ldots k_n, \epsilon_{\mu_n}(k_n)|S|p, s\rangle, \tag{4.26}$$

where p, s and p', s' denote the initial and final electron momenta and spins, and the k_n with polarization index ϵ_{μ_n} describe the emitted photons. The states are represented by

$$|p, s\rangle = b_s^+(p)\rangle$$

and

$$\begin{aligned}\langle p', s'; k_1\epsilon_1; \ldots k_n\epsilon_n| = \frac{1}{\sqrt{n!}}\epsilon_{\mu_1}(k_1)\ldots\epsilon_{\mu_n}(k_n)\\ \cdot\langle b_{s'}(p')\cdot a_{\mu_1}(k_1)\ldots a_{\mu_n}(k_n),\end{aligned} \tag{4.27}$$

with the IN-field operators comprising these states commuting through the Wick-ordered brackets of (4.25). Each time this occurs, the factors

$$\int dz[a_{\mu_i}(k_i), A_\nu^{IN}(z)]\vec{K}_z\frac{\delta}{\delta j_\nu(z)}$$

and

$$\sum_{\alpha,\beta}\int d^4x'\{b_{s'}(p'), \bar{\psi}_\alpha^{IN}(x')\}(\vec{\mathcal{D}}_{x'})_{\alpha\beta}\frac{\delta}{\delta\bar{\eta}_\beta(x')}$$

---

[4] See any text on quantum field theory for the renormalization procedures, for example any of [Bjorken and Drell (1965)]; [Bogoluibov and Shirkov (1976)]; [Itzykson and Zuber (1980)]. In the present functional context, this is exhibited in Chapter 6 of [Fried (1972)].

and

$$-\sum_{\alpha\beta}\int d^4x \frac{\delta}{\delta\eta_\alpha(x)}\left(\overleftarrow{\bar{\mathcal{D}}}_x\right)_{\alpha\beta}\{\psi_\beta^{IN}(x), b_s^+(p)\} \tag{4.28}$$

will appear, with the FDs operating upon the GF. From (4.28) and the free-particle commutation and anticommutation relations, one easily obtains for (4.26),

$$\begin{aligned}
-(2\pi)^{-\frac{3}{2}(n+2)} &\cdot \frac{1}{\sqrt{n!}}\sum_{\mu_1}\epsilon_{\mu_1}(k_1)\cdots\sum_{\mu_n}\epsilon_{\mu_n}(k_n)\cdot\left[\frac{m^2}{EE'}\right]^{1/2}\\
&\cdot (2\omega_1\ldots 2\omega_n)^{-1/2}\cdot\int d^4x' e^{-ip'\cdot x'}\bar{u}_{s'}^{\alpha'}(\vec{\mathcal{D}}_{x'})_{\alpha'\beta'}\\
&\cdot\int d^4x e^{ip\cdot x}(\vec{\bar{\mathcal{D}}}_x)_{\alpha\beta}u_s^\beta(p)\\
&\cdot\langle S\rangle\cdot\int d^4z_1\ldots\int d^4z_n e^{-i(k_1\cdot z_1+\cdots k_n\cdot z_n)}\\
&\cdot\vec{K}_{z_1}\ldots\vec{K}_{z_n}\frac{\delta}{\delta j_{\mu_1}(z_1)}\cdots\frac{\delta}{\delta j_{\mu_n}(z_n)}\frac{\delta}{\delta\bar{\eta}_{\beta'}(x')}\\
&\cdot\frac{\delta}{\delta\eta_\alpha(x)}\cdot Z\{j,\eta,\bar{\eta}\}|_0.
\end{aligned} \tag{4.29}$$

From this one sees that the passage from GF to S-matrix element involves the calculation of a FD of Z for every asymptotic particle in the problem; the operation upon that quantity by Dirac or Klein-Gordon operators, as appropriate; and, finally, the Fourier transform of all configuration-space dependence in terms of the appropriate mass-shell momenta (plus spins and polarizations). These last two steps are frequently referred to as "mass-shell amputation".

Chapter 5

Schwinger/Fradkin Representations

In every causal theory, relativistic or non-relativistic, there is a way to phrase the physical quantities of interest in terms of Green's functions defined in the presence of some "background" field. For problems of potential theory, or first quantization, one needs to know or approximate Green's functions such as the $G_c(x, y|A)$ of Chapter 3, with the form of $A(z)$ specified by the particular problem under study. In contrast, second quantization demands the calculation of quantum fluctuations given by summing over Gaussian-weighted fluctuations of the A-dependence of products of such $G_c[A]$, together with appropriate determinant factors, $\exp\{L[A]\}$. If the first step in approaching the calculation of any physical quantity is its functional representation in terms of $G_c[A]$, the second is surely a representation of such Green's functions in as explicit and as useful a manner as possible.

In this Chapter we describe the generic, proper-time representations invented by Schwinger [Schwinger (1951)] and made explicit by Fradkin [Fradkin (1966)], and explain why the latter are so convenient for QFT. Svidinsky's simplified variant of the representation for $G_c[A]$, useful in eikonal approximations to fermion theories, is constructed. In this Chapter, we will hold to the $G_c[A]$ of QED and (if $A_\mu \to A_\mu^a \lambda_a$, with λ_a the Gell-Mann, or defining-representations matrices of SU(N)) of QCD.

5.1 Formalism

The basic differential equation for $(QED)_4$ is

$$(m + \gamma_\mu[\partial_\mu - igA_\mu(x)])G_c(x, y|A) = \delta^4(x - y), \tag{5.1}$$

which can be rewritten in a symbolic, operator form (for any dimension) as

$$G_c[A] = (m - i\gamma \cdot \Pi)^{-1}, \tag{5.2}$$

where $\Pi_\mu = i(\partial_\mu - igA_\mu)$, and the operators ∂ and A are local quantities in the sense that $\langle x|\partial|y\rangle = \partial^x \delta(x-y)$ and $\langle x|A|y\rangle = A(x) \cdot \delta(x-y)$; in this way, $\langle x|G_c[A]|y\rangle = G_c(x,y|A)$. We will use g everywhere to denote the fermionic electric (and, in principle, unrenormalized) charge. When $g \to 0, G_c \to S_c$, the free-fermion propagator of (1.19), and this permits one to write an integral equation for G_c in the form

$$G_c = S_c + igS_c(\gamma \cdot A)G_c, \tag{5.3}$$

or

$$G_c = S_c + igG_c(\gamma \cdot A)S_c, \tag{5.4}$$

using the convenient, formal notation. Equations(5.3) and (5.4) are equivalent in so far as their perturbative expansions are concerned; but for strong coupling problems, these are not of overly great value.

In 1951 Schwinger introduced a formal, "proper-time" representation for G_c, in the following way. One first "rationalizes" (4.2) by rewriting it as

$$G_c = (m + i\gamma \cdot \Pi) \cdot [(m - i\gamma \cdot \Pi)^{\cdot}(m + i\gamma \cdot \Pi)]^{-1},$$

or

$$G_c = (m + i\gamma \cdot \Pi)^{\cdot}[m^2 + (\gamma \cdot \Pi)^2]^{-1}, \tag{5.5}$$

where: $(\gamma \cdot \Pi)^2 = \Pi^2 + ig(\sigma \cdot F)$, with $\sigma_{\mu\nu} = (1/4)[\gamma_\mu, \gamma_\nu]$. Remembering that m is to have an infinitesimal, negative imaginary part, as appropriate to the definition of this causal propagator, one introduces the representation

$$[m^2 + (\gamma \cdot \Pi)^2]^{-1} = i \int_0^\infty ds \exp\left\{-is[m^2 + (\gamma \cdot \Pi)^2]\right\}, \tag{5.6}$$

where the "proper-time" variable s really has the dimensions of (time)^2; subsequently, the continuation $s \to i\tau$ will be made, and τ will be then referred to as the proper time.

Before discussing Fradkin's representation of $\exp\{-(\gamma \cdot \Pi)^2\}$, it will be useful to note Schwinger's representation for the fermion closed-loop functional $L[A] = Tr \ln(1 - ig(\gamma \cdot A)S_c)$, quite similar to that of (5.6). Using the parametric representation of (3.25),

$$L[A] = -i \int_0^g dg' Tr \left\{(\gamma \cdot A)S_c[1 - ig'(\gamma \cdot A)S_c]^{-1}\right\},$$

or

$$L[A] = -i \int_0^g dg' Tr\{(\gamma \cdot A)G_c[g'A]\}, \tag{5.7}$$

one substitutes into (5.7) the form (5.5) and the representation (5.6), and discards all terms proportional to the vanishing (Dirac) trace over an odd number of γs to obtain

$$\begin{aligned} L[A] = i &\int_0^\infty ds \exp[-ism^2] \\ &\cdot \int_0^g dg' Tr \left\{(\gamma \cdot A)(\gamma \cdot \Pi) \exp[-is(\gamma \cdot \Pi)^2]\right\}, \end{aligned} \tag{5.8}$$

where the coupling constant inside Π is g'. Because of the Tr trace operations, (5.8) can be rewritten as

$$\begin{aligned} L[A] = -\frac{1}{2} &\int_0^\infty ds s^{-1} \exp[-ism^2] \\ &\cdot \int_0^g dg' \frac{\partial}{\partial g'} Tr \left\{\exp[-is(\gamma \cdot \Pi)^2]\right\}, \end{aligned}$$

or

$$L[A] = -\frac{1}{2} \int_0^\infty ds s^{-1} \exp[-ism^2] Tr \left\{\exp[-is(\gamma \cdot \Pi)^2]\right\} - \{g = 0\}, \tag{5.9}$$

where the coupling constant of Π is again g.

For both $G_c[A]$ and $L[A]$, the essential quantity to be understood is $U(s) = \exp[-is(\gamma \cdot \Pi)^2]$. A perturbative development, along with a solution for the two special cases of constant $F_{\mu\nu}$ and fields depending upon a single frequency, was given by Schwinger; but it is possible to find a non-perturbative development with the aid of the representation introduced by Fradkin. For this, one replaces $U(s)$ by the seemingly more complicated quantity, the ordered exponential

$$U(s, v) = (\exp\{-i \int_0^s ds' [\Pi^2 + ig\sigma \cdot F + v_\mu(s')\Pi_\mu]\})_+, \tag{5.10}$$

with the property that $U(s, v)|_{v \to 0} = U(s)$. One notes that $U(s, v)$ satisfies the relations

$$\frac{\partial U}{\partial s} = -i[\Pi^2 + ig\sigma \cdot F + v(s) \cdot \Pi]U \tag{5.11}$$

and

$$\delta U / \delta v_\mu(s) = -i\Pi_\mu U \tag{5.12}$$

and the initial condition $U|_{s=o} = 1$.

The reason for doing this is that $U(s, v)$ can be given a particularly elegant representation, in the form

$$U = \exp\left\{i \int_0^{s-\epsilon} ds' \delta^2 / \delta v_\mu^2(s')\right\} W(s, v)|_{\epsilon \to 0}$$

with

$$W(s, v) = \left(\exp\left\{-i \int_0^s ds' [v_\mu(s')\Pi_\mu + ig\sigma \cdot F]\right\}\right)_+. \tag{5.13}$$

It is easy to see that (5.13) provides a solution to (5.11) and (5.12) in the limit $\epsilon \to 0$, and we shall assume that limit in everything that follows. The importance of (5.13) is that W may be given a representation in which the field dependence is made explicit, with the result that $L[A]$, or $G_c[A]$, may be expressed exactly in terms of the Gaussian fluctuations (over the values of $v(s')$) of that comparatively simple W. The form of those fluctuations will be given below in terms of a Gaussian functional integral.

We now calculate, explicitly, configuration-space matrix elements of W, and begin by writing the LHS projection of W as

$$\begin{aligned} \langle x|W(s) &= \langle x| \exp\left\{\int_0^s ds' v_\mu(s')\partial_\mu\right\} \mathcal{F}(s) \\ &= \exp\left\{\int_0^s ds' v \cdot \partial_x\right\} \langle x|\mathcal{F}(s). \end{aligned} \tag{5.14}$$

Using the differential equation satisfied by $W(s)$, which may be read off from its definition in (5.13), one finds

$$\left(\frac{\partial}{\partial s}\right)\langle x|\mathcal{F}(s) = -ig\exp\left\{-\int_0^s ds'v\cdot\partial_x\right\}[v(s)\cdot A(x) + i\sigma\cdot\mathcal{F}(x)]$$
$$\cdot\exp\left\{\int_0^s ds'v\cdot\partial_x\right\}\langle x|\mathcal{F}(s)$$
$$= -ig\big[v(s)\cdot A(x - \int_0^s ds'v(s'))$$
$$+g\sigma\cdot F(x - \int_0^s ds'v(s'))\big]\langle x|\mathcal{F}(s), \tag{5.15}$$

which is an explicit equation for $\langle x|\mathcal{F}(s)$ in terms of the fields, and has as its solution

$$\langle x|F(s) = \Bigg(\exp[-ig\int_0^s ds'\{v(s')\cdot A(x - \int_0^{s'} ds''v(s''))$$
$$+g\sigma\cdot F(x - \int_0^{s'} ds''v(s''))\}]\Bigg)_+\langle x|,$$

so that

$$\langle x|W(s)|y\rangle =$$
$$\left(\exp\left[-ig\int_0^s ds'\{v(s')\cdot A(y - \int_0^{s'} v) + g\sigma\cdot F(y - \int_0^{s'} v)\}\right]\right)_+$$
$$\cdot\quad\delta(x - y + \int_0^s v). \tag{5.16}$$

Note that the ordered bracket is necessary in passing from (5.15) to (5.16) only because of the s'-dependent (Dirac) matrix $\sigma\cdot F$. In two dimensions, however, there is just one antisymmetric matrix σ_{14} which can enter, while for a constant $F_{\mu\nu}$, $\sigma\cdot F$ becomes a constant (independent of s') matrix which commutes with itself and unity; in both cases, therefore, the ordered exponential (OE) becomes an ordinary exponential involving either σ_{14} or the constant $\sigma\cdot F$. The general OE of (5.16) has the same interpretation as the operator brackets displayed in Chapters 1 and 3, except that the ordering symbol here means an ordering with respect to the s' variable: for any expansion of the exponent, those quantities carrying the larger values of s' are to stand furthest to the left. A general discussion of OEs , and two strong-coupling methods for their approximation, may be found in *Green's Functions and Ordered Exponentials* [Fried (2002)].

With the aid of (5.5),(5.6), and the Fradkin solution for (5.13), one can write, finally, the representations

$$G_c(x,y|A) = i\int_0^\infty ds e^{-ism^2} \cdot \exp\left[i\int_0^s ds' \frac{\delta^2}{\delta v^2(s')}\right] \cdot \left(m - \gamma_\mu \frac{\delta}{\delta v_\mu(s)}\right)\langle x|W(s)|y\rangle|_{v\to 0}, \tag{5.17}$$

and

$$L[A] = -\frac{1}{2}\int_0^\infty \frac{ds}{s} e^{-ism^2} \cdot tr\int d^D x \cdot \exp\left[i\int_0^s ds' \frac{\delta^2}{\delta v^2(s')}\right] \cdot \left\{\langle x|W(s)|x\rangle - \langle x|W(s)|x\rangle|_{g=0}\right\}|_{v\to 0}, \tag{5.18}$$

where tr denotes a trace over Dirac coordinates, and in which all the field dependence is explicit (except that of the $\sigma \cdot F$ factor inside the OE).

5.2 Gauge Structure

In order to isolate the gauge-variant part of $\langle x|W|y\rangle$ — that is, those parts which change under the $U(1)$ gauge transformation of QED, $A_\mu \to A'_\mu = A_\mu + \partial_\mu \Lambda$ – it is useful to consider the quantity

$$Q(\lambda) \equiv \int_0^s ds' v_\mu(s') A_\mu(y - \lambda\int_0^{s'} v),$$

which for $\lambda = 1$ appears in the exponential of (5.16). A simple integration-by-parts produces

$$Q(\lambda) = A_\mu(y - \lambda\int_0^s v) \cdot \int_0^s ds' v_\mu(s') + \lambda\int_0^s ds' \int_0^{s'} ds'' v_\mu(s'') v_\nu(s') \partial_\nu A_\mu(y - \lambda\int_0^{s'} v), \tag{5.19}$$

but, because of the $\delta(x - y + \int_0^s v)$ factor of (5.16) the first RHS term of (5.19) may be replaced by $(y-x)_\mu A_\mu(y - \lambda(y-x))$. Then, with the definition of $F_{\nu\mu}$, (5.19) can be rewritten as

$$Q(\lambda) = (y-x)_\mu A_\mu(\lambda x + (1-\lambda)y)$$
$$+ \lambda \int_0^s ds' \int_0^{s'} ds'' v_\mu(s'') v_\nu(s')$$
$$\cdot \left[F_{\mu\nu}(y - \lambda \int_0^{s'} v) + \partial_\mu A_\nu(y - \lambda \int_0^{s'} v)\right]. \quad (5.20)$$

But the last RHS term of (5.20) may be replaced by

$$-\lambda \int_0^s ds' v_\nu(s') \frac{\partial}{\partial\lambda} A_\nu(y - \lambda \int_0^{s'} v) = -\lambda \frac{\partial}{\partial\lambda} Q(\lambda),$$

which, when substituted into (5.20), produces the differential equation

$$\frac{\partial}{\partial\lambda}(\lambda Q(\lambda)) = (y-x)_\mu A_\mu(\lambda x + (1-\lambda)y)$$
$$+ \lambda \int_0^s ds' \int_0^{s'} ds'' v_\mu(s') v_\nu(s'') F_{\mu\nu}(y - \lambda \int_0^{s'} v). \quad (5.21)$$

The integral of (5.21) between $\lambda = 0$ and $\lambda = 1$ is immediate and yields

$$Q(1) = -\int_y^x d\xi_\mu A_\mu(\xi) + \int_0^1 \lambda d\lambda \int_0^s ds'$$
$$\cdot \int_0^{s'} ds'' v_\mu(s') v_\nu(s'') F_{\mu\nu}(y - \lambda \int_0^{s'} v), \quad (5.22)$$

where ξ_μ denotes the straight-line path between x_μ and y_μ, $\xi_\mu = \lambda x_\mu + (1-\lambda) y_\mu$.

By this simple computation, one sees that $L[A]$ is gauge invariant; and that the only gauge-variant part of $G_c(x, y|A\}$ is that factor coming from (5.22), $\exp[ig \int d\xi_\mu A_\mu(\xi)]$. Under the gauge change $A_\mu \to A_\mu + \partial_\mu \Lambda$, the only variation of the complete G_c is

$$G_c(x, y|A + \partial\Lambda) = e^{ig[\Lambda(x) - \Lambda(y)]} G_c(x, y|A), \quad (5.23)$$

a property which may also be inferred directly from the equations which define $G_c[A]$. As noted elsewhere[1] properly defined S-matrix elements are independent of such gauge changes.

[1] [Fried (1972)], Chapter 6

It will also be useful to comment on the corresponding gauge properties found in QCD, with the definitions of $G_c[A]$ and $L[A]$ the same as in QED except for the replacement $A_\mu \to A_\mu^a \lambda^a$ where the λ^a are the the fundamental or defining matrix representations of SU(N) (the Gell-Mann matrices) satisfying

$$\begin{aligned} [\lambda^a, \lambda^b] = 2if_{abc}\lambda^c, \ \{\lambda^a, \lambda^b\} = \frac{4}{N}\delta_{ab} + 2d_{abc}\lambda^c, \\ tr[\lambda^a] = 0, \ tr[\lambda^a\lambda^b] = 2\delta_{ab}. \end{aligned} \tag{5.24}$$

Following from an original Lagrangian density of form

$$\mathcal{L} = -\frac{1}{4}(F_{\mu\nu}^a)^2 - \bar{\psi}[m + \gamma_\mu(\partial_\mu - igA_\mu^a\lambda^a)]\psi,$$

local (that is, position-dependent) gauge transformations which leave this quark-gluon Lagrangian invariant are defined by

$$\begin{aligned} A_\mu^a(z)\lambda^a \equiv A_\mu(z) \to A'_\mu(z) = V^+(z)(A_\mu(z) + \tfrac{i}{g}\partial_\mu)V(z), \\ F_{\mu\nu}^a(z)\lambda^a \equiv F_{\mu\nu}(z) \to F'_{\mu\nu}(z) = V^+(z)F_{\mu\nu}(z)V(z), \end{aligned} \tag{5.25}$$

for arbitrary $V(z) = \exp[i\lambda^a\omega^a]$, where

$$F_{\mu\nu}^a = \partial_\mu A_\mu^a - \partial_\nu A_\mu^a + gf_{abc}A_\mu^b A_\nu^c.$$

The Fradkin representations for $G_c[A]$ and $L[A]$ go through as before, except that the color matrix factors λ^a require the use of OEs everywhere in the formula corresponding to (5.16), and the tr operation includes a summation over such color variables.

The invariance of $L[A]$ under the full transformation (5.25) can be shown in a simple way, by writing $\langle x|W(s)|x\rangle = U(x)\delta(\int_0^s ds'v(s'))$ and seeing how $U(s)$ changes under a gauge transformation. Under such a transformation, $U(s) \to U'(s)$ where

$$\begin{aligned} U'(s) = \Big(\exp\Big[-ig\int_0^s ds'\{v_\mu(s')V^+(x - \int_0^{s'} v)(A_\mu + \frac{i}{g}\partial_\mu)V - i\sigma_{\mu\nu} \\ \cdot V^+(x - \int_0^{s'} v)F_{\mu\nu}V\}\Big]\Big)_+, \end{aligned} \tag{5.26}$$

and where, in writing (5.26) and similar expressions, dependence on the common variable $x - \int_0^{s'} ds''v(s'')$ is exhibited only in the first term of any product. To understand the relation between U and U' it is useful to consider the differential equation for U',

$$\frac{\partial U'}{\partial s} = -ig[v_\mu(s)V^+(x - \int_0^s v)(A_\mu + \frac{i}{g}\partial_\mu)V - i\sigma_{\mu\nu}V^+(x - \int_0^s v)F_{\mu\nu}V] \cdot U'. \tag{5.27}$$

Setting $U' = V^+(s)Z(s)$, with $V(s) \equiv V(x - \int_0^s v)$, substitution into (5.27) builds the equation for $Z(s)$,

$$\frac{\partial Z}{\partial s} = -ig\left[v_\mu(s)A_\mu(x - \int_0^s v) - i\sigma_{\mu\nu}F_{\mu\nu}(x - \int_0^s v)\right] \cdot Z, \tag{5.28}$$

where the replacements $v_\mu(s)V^+(s)\partial_\mu V(s) = -V^+(s)\frac{\partial V}{\partial s}$ and $\frac{\partial V^+}{\partial s} \cdot V = -V^+\frac{\partial V}{\partial s}$ have been made.

Taking into account the initial condition $Z(0) = V(0)$, and in comparison with the equation and solution for $U(s)$, one can write the solution of (5.28) as $Z(s) = U(s)V(0)$, so that, finally, $U'(s) = V^+(s)U(s)V(0)$. But $V^+(s) = V^+(0)$, if the closed loop condition $\int_0^s ds'v(s') = 0$ is satisfied, as required by the representation (5.18). Hence $tr[U] = tr[U']$, and $L[A]$ has been explicitly shown to be invariant under the full gauge transformations of QCD. This is not a surprise, of course, for the Fradkin representation is exact; but it is useful to see how the exact gauge invariance is fulfilled before attempting any approximations to this closed-loop functional.

5.3 The Bloch–Nordsieck - IR - Eikonal Approximation

The first step in any evaluation of these forms is to introduce a Fourier representation for the $\delta(x - y - \int_0^s ds'v(s'))$ factor, which contributes a term linear in $v(s')$ to the exponential of (5.16). One can see at a glance why and how the special case of constant $F_{\mu\nu}$ leads to an exact solution, for the exponent of $\langle x|W|y\rangle$ is then quadratic in $v(s')$, so that the functional operations are Gaussian and can be performed immediately.

It is sometimes convenient to change to a FI over fluctuations of the $v(s')$, and following the discussion of Chapter 2, Section 2.4, this can be done by breaking up the parameter range 0 to s into small intervals in the manner of (2.16), so that one finds, for example in QED

$$L[F] = -\frac{1}{2}\int_0^\infty \frac{ds}{s} e^{-ism^2} \int \frac{d^{\mathcal{D}}p}{(2\pi)^{\mathcal{D}}} \int d^{\mathcal{D}}x \cdot N(s) \cdot \int d[\phi_\mu]$$
$$\cdot \exp[\frac{i}{4}\int_0^s ds'\phi^2(s') + i\int_0^s ds' p_\nu \phi_\nu(s')]$$
$$\cdot\, tr(\exp[g\int_0^s ds'\sigma \cdot F(x - \int_0^{s'} \phi])_+$$
$$\cdot \exp[-ig\int_0^s ds' \int_0^{s'} ds'' \phi_\mu(s')\phi_\nu(s'')$$
$$\cdot \int_0^1 \lambda d\lambda F_{\mu\nu}(x - \lambda \int_0^{s'} \phi)],$$

where the LHS of (5.29) has been written as $L[F]$ instead of as $L[A]$, to emphasize the manifest gauge invariance of this closed-loop functional; and where the normalization constant $N(s)$ is given by

$$N(s)^{-1} = \int d[\phi] \exp\,[\frac{i}{4}\int_0^s ds'\phi^2(s')].$$

In the passing from (5.18) to (5.29), one considers fluctuations over the s'-dependent $\phi_\mu(s')$, instead of the $\nu_\mu(s')$; but the important (computational) point is that these exact representations for $G_c[A]$ and $L[A]$ are given in terms of Gaussian-weighted fluctuations over field dependence which is essentially explicit. (In scalar theories, without spin or color degrees of freedom, this is precisely true.) If the $F_{\mu\nu}$ are constant, the fluctuation FI is Gaussian and can be performed exactly, leading to the Schwinger, constant-field solutions.

If the $F_{\mu\nu}$ are not constants, but if there is reason to extract only the low-frequency parts of the photon or gluon interactions with fermions, then one form of a "natural" Bloch-Nordsieck (BN) approximation is to replace the ϕ_μ dependence inside the $F_{\mu\nu}$ by a constant 4-vector ϕ^0_μ, which corresponds in BN theory (below) to the 4-velocity of an incident or final scattering fermion. With that identification, the FI for $G_c[A]$ is still Gaussian and can be performed exactly; in this way, in principle, one can reproduce all the eikonal approximations ever written for scattering and production amplitudes. (There are easier ways of accomplishing this, as noted below; but none which so clearly shows the relation between the exact and approximate forms.)

Finally, it may be noted that all of conventional perturbation theory can be encompassed by the remark that any expansion of (5.17) or (5.18) in powers of the coupling constant g will, in any order of approximation, again produce a (different) Gaussian FI; in (5.29), e.g., every perturbation term is just a FI over a polynomial in $\phi_\mu(s'')$ weighted by the Gaussian factor $\exp[(i/4)\int ds'\phi^2]$.

The only approximation discussed in this Chapter is Svidinski's construction [Symanzik (1960)]; [Svidinsky and Eksper (1956)] of the fermion Green's function used in conventional BN approximations. One begins with the differential equation (5.1) for $G_c(x,y|A)$, and imagines a scattering process in which all real and virtual photons are restricted to very small frequencies, that is, to energies much less than the asymptotic energies of the scattering particles. Then, the recoil of the fermion emitting or absorbing such "soft" photons is negligible, which suggests that the Dirac γ_μ in (5.1) may be replaced by an "average value", $-\,iv_\mu$, where v_μ represents the 4-velocity of an incident or outgoing mass-shell fermion, $v_\mu = p_\mu/m$, $v^2 = -1$. (This is the step that corresponds, in the Fradkin representation for $G_c[A]$, to the replacement of the $v_\mu(s')$ in the arguments of A_μ or $F_{\mu\nu}$ of (5.16) by a constant 4-velocity associated with asymptotic momenta.) While useful in setting up eikonal models, one must not take this approximation seriously if one is calculating anything sensitive to high-frequency components, such as perturbative renormalization constants or mass shifts.

One may then examine the new equation resulting from this approximation,

$$[m - iv\cdot(\partial - igA)]G_{BN}(x,y|A) = \delta^4(x-y),$$

or

$$G_{BN}[A] = [m - iv\cdot(\partial - igA)]^{1} \tag{5.29}$$

in the compact, formal notation. Because (5.29) specifies a single, first-order differential equation, its solution will be either retarded or advanced, rather than one of the four possibilities (advanced, retarded, causal, anti-causal) of the ordinary, relativistic Dirac (or second-order scalar) equation. Use of the prescription $m \to m - i\epsilon$ produces a retarded function for the BN solution $G_{BN}(x,y|A)$, in terms of the proper-time representation

$$G_{BN}[A] = i\int_0^\infty ds e^{-ism}\cdot e^{-sv\cdot(\partial - igA)},$$

or

$$G_{BN}[A] = i\int_0^\infty dse^{-ism}\cdot e^{-sv\cdot\partial}\mathcal{F}(s),$$

with

$$\mathcal{F}(s) = e^{sv\cdot\partial}\cdot e^{-sv\cdot(\partial - igA)}.$$

More simply than in the computation of (5.14), one can construct and solve an equation for $\mathcal{F}(s)$,

$$\frac{\partial}{\partial s}\langle x|\mathcal{F}(s) = igv\cdot A(x+sv)\langle x|\mathcal{F}(s),$$

with solution

$$\langle x|\mathcal{F}(s) = e^{ig\int_0^s ds' v\cdot A(x+s'v)}\langle x|,$$

so that

$$G_{BN}(x,y|A) = i\int_0^\infty dse^{-ism}\delta(x-y-sv)\cdot e^{ig\int_0^s ds' v\cdot A(y+s'v)}. \tag{5.30}$$

Note that, because $v_0 = E/m > 0$, this G_{BN} is a retarded function; and hence, any closed-fermion-loop $L[A]$ constructed from this G_{BN} must vanish.

For use in subsequent eikonal calculations, we note here the result of mass-shell amputation applied to this BN propagator; that is, operating on the LHS x-variable, one calculates

$$\int d^4xe^{-ip\cdot x}(m - iv\cdot\vec{\partial}_x)G_{BN}(x,y|A)|_{m+v\cdot p=0}. \tag{5.31}$$

Before the mass-shell limit is taken, (5.31) can be rewritten as

$$i(m+v\cdot p)e^{-ip\cdot y}\int_0^\infty ds\exp\big[-is(m+v\cdot p) + ig\int_0^s ds' v\cdot A(y+s'v)\big]$$

or as

$$-e^{-ip\cdot y}\int_0^\infty ds\exp\big[ig\int_0^s ds' v\cdot A(y+s'v)\big]\cdot\frac{\partial}{\partial s}e^{-is(m+v\cdot p)}$$

and, after an integration-by-parts, as

$$e^{-ip\cdot y}\{1+\int_0^\infty dse^{-is(m+v\cdot p)}\cdot\frac{\partial}{\partial s}e^{ig\int_0^s ds'v\cdot A(y+s'v)}\}.$$

Now the mass-shell limit may be taken, $m+v\cdot p\to 0$ (which is appropriate if $v_\mu = p_\mu/m$), and one finds for (5.31) simply

$$e^{-ip\cdot y}\cdot e^{ig\int_0^\infty dsv\cdot A(y+sv)}. \tag{5.32}$$

In the same manner, mass-shell amputation with respect to the RHS y-variable of G_{BN} leads to

$$\int d^4ye^{ip'\cdot y}G_{BN}(x,y|A)(m+v'\cdot p')|_{m+v'\cdot p'=0}$$
$$= e^{ip'\cdot x}\cdot e^{ig\int_0^\infty dsv'\cdot A(x-sv')}. \tag{5.33}$$

5.4 A Convenient Reformulation

A most convenient reformulation of the representations of (5.17) and (5.18) is obtained by employing the new variable $u_\mu(s')=\int_0^{s'}ds''v_\mu(s'')$; that is, instead of using Fradkin's four-velocity to describe the virtual motion of the fermion, in its interactions with real or virtual photons (or, in QCD, with gluons), one changes to the space-time coordinates $u_\mu(s')$. Consider, for example, the complete expression of (5.18),

$$\begin{aligned}L[A]=&-\frac{1}{2}\int_0^\infty\frac{ds}{s}e^{-ism^2}\cdot tr\int d^4x\exp\Big[i\int_0^s ds'\frac{\delta^2}{\delta v^2(s')}\Big]\\&\cdot\delta^{(4)}\Big(\int_0^s ds'v(s')\Big)\cdot\Big(\exp\Big[-ig\int_0^s ds'\{v(s')\cdot A(y-\int_0^{s'}v)\\&+i\sigma\cdot F(y-\int_0^{s'}v)\}\Big]\Big)_+-(g=0),\end{aligned} \tag{5.34}$$

and under all of the integrals of (5.34), insert a factor of unity, written in the form

$$1=\int d[u]\delta^{(4)}\Big[u(s')-\int_0^{s'}ds''v(s'')\Big], \tag{5.35}$$

so that the replacements

$$\delta\Big(\int_0^s v\Big) \to \delta\big(u(s)\big), \ v(s')\cdot A\Big(y - \int_0^{s'} v\Big) \to u'(s')\cdot A(y - u(s')) \tag{5.36}$$

may be made in (5.34). Then, write a representation for the delta functional of (5.35),

$$\delta^{(4)}\left[u(s') - \int_0^{s'} v\right] = N' \int d[\phi]\cdot \exp\left[i\int_0^s ds'\phi_\mu(s')\Big(u_\mu(s') - \int_0^{s'} ds'' v_\mu(s'')\Big)\right],$$

where N' is a normalization constant defined below, and the FIs $\int d[u]$ and $\int d[\phi]$ have been defined in Chapter 2. Abel's trick can now be used to rewrite the second exponential factor on the RHS of (5.37) as

$$-i\int_0^s ds' v_\mu(s') \int_{s'}^s ds''\phi_\mu(s''),$$

and after the replacements (5.36), the only remaining v-dependence of (5.34) then takes the form

$$\begin{aligned} & e^{i\int_0^s \frac{\delta^2}{\delta v^2}} \cdot e^{-i\int_0^s v(s')ds' \int_0^s ds''\theta(s''-s')\phi(s'')}\Big|_{v\to 0} \\ &= \exp\left[-i\int_0^s ds'\Big(\int_0^s ds''\theta(s''-s')\,\phi(s'')\Big)^2\right] \\ &= \exp\left[-i\int_0^s ds_1\int_0^s ds_2\ \phi(s_1)\cdot\phi(s_2)\cdot h(s_1,s_2)\right], \end{aligned} \tag{5.37}$$

where $h(s_1,s_2) = \int_0^s ds'\theta(s_1-s')\theta(s_2-s') = \theta(s_1-s_2)s_2 + \theta(s_2-s_1)s_1 = \frac{1}{2}\big[(s_1+s_2) - |s_1-s_2|\big]$.

The FI over ϕ is now simply Gaussian, and can be read off from (2.28),

$$N'\int d[\phi]e^{-i\int\phi\cdot h\cdot\phi + i\int\phi\cdot u} = N'\cdot C' e^{-\frac{1}{2}Tr\ln[2h]}\cdot e^{\frac{i}{4}\int u[2h]^{-1}u}.$$

Inspection of (5.37) shows that the constant $N' = \frac{\pi}{i}(\frac{\Delta}{2\pi^1})$, using the same notation for the product of integration over each of the cells denoted by the subscript i. Combining this with the C' obtained from (2.28), $C' = \frac{\pi}{i}(\frac{2\pi}{i\Delta})^{\frac{1}{2}}$, the product $N'C'$ is then: $\frac{\pi}{i}(\frac{\Delta}{2\pi i})^{\frac{1}{2}}$. When the Gaussian-weighted FI $\int d[u]$

is performed over the remaining u-dependence — which is itself trivially[2] Gaussian (after the extraction of the F-dependence from its ordered exponential) — there will result a new constant, C'', multiplying $N'C'$, where $C'' = \frac{\pi}{i}(\frac{2\pi i}{\Delta})^{\frac{1}{2}} = C$ of (2.28); so that $N'C'C'' = 1$. In this way, all such normalization constants, including the factor $\exp\{-(1/2)Tr\ln(2h)\}$, are exactly removed upon calculating the FI over u.

Writing $N'C' = C^{-1}$, the $L[A]$ of QED has the Fradkin representation

$$L[A] = -\frac{1}{2}\int_0^\infty \frac{ds}{s}e^{-ism^2}\cdot\int d^4x\cdot tr\cdot C^{-1}e^{-\frac{1}{2}Tr\ln[2h]} \cdot\int d[u]e^{\frac{1}{2}\int u[2h]^{-1}u}\cdot\delta^{(4)}(u(s)) \cdot\left\{e^{-is\int_0^s ds'u'(s')\cdot A(y-u(s'))}\left(e^{g\int_0^s ds'\sigma\cdot F(x-u(s'))}\right)_+ - (g=0)\right. \tag{5.38}$$

In a similar fashion, there appears the more convenient QED expression for $G_c(x,y|A)$,

$$G_c(x,y|A) = i\int_0^\infty dse^{-ism^2}\cdot\left(c^{-1}\cdot e^{-\frac{1}{2}Tr\ln[2h]}\right) \cdot\int d[u]e^{\frac{1}{2}\int u(2h)^{-1}u}\cdot\delta^{(4)}(x-y+u(s')) \cdot e^{-ig\int_0^s ds'u'(s')\cdot\, A(y-u(s'))}\left(e^{g\int_0^s ds'\sigma\cdot F(y-u(s'))}\right)_+ . \tag{5.39}$$

One can now anticipate the qualitative results that will follow, in both QED and QCD, because — ordered or not — the A-dependence of these integrands is Gaussian. This means that the linkage operations acting upon the A-dependence can be carried through exactly, and this corresponds to performing non-perturbative sums over all relevant Feynman graphs, expressing their results in terms of the Fradkin functional representations.

The immediate question is then: "Is this really progress?" And the answer is positive, simply because the Fradkin representations are Potential Theory constructs, and there exists simple approximations to extract the information they carry. For example, in High-Energy Scattering, the Fradkin representation for $G_c[A]$ goes over into a Bloch-Nordsieck/eikonal representation, which is simplicity itself. Further, there are functional tricks which

[2]By "trivially" is meant that the QED A-dependence will appear linearly in the exponent of the result of the FI. In QCD, for reasons of color-gauge invariance as explained in Part 3, there will be an additional factor multiplying these representations, which will contain an exponential of quadratic A-dependence. Hence the linkage integrands will again be Gaussian, and the functional operations can again be carried through exactly.

can be used to simplify $L[A]$, as will appear in the estimations described below, for QED and QCD.

In summary, employing Fradkin representations for $G_c[A]$ and $L[A]$ in the Schwinger/Symanzik functional representations for the GFs of QED and QCD, permits a direct entry into non-perturbative QFT, independently of the strength of the coupling. And when this approach is followed there appear surprises, both in QED and QCD.

PART 2
Quantum Electrodynamics

Quantum Electrodynamics

QED is a vast subject, with many facets, each with multiple subsections. The use of Schwinger/Symanzik/Fradkin functional methods is a most efficient way to set up and complete specific calculations with a minimum of irrelevant distractions: Gauge-invariance of radiative corrections to photonic processes is always guaranteed, while mass-shell amputation removes gauge ambiguities for processes involving charged fermions. Infrared (IR) divergences found in specific perturbative calculations are automatically, and easily, removed with the aid of a relevant, non-perturbative Bloch-Nordsieck (BN) approximation, while a simple, functional observation removes the need for tedious cancellations when computing sums of higher-order Feynman graphs.

Many of the above topics have already been discussed, from a functional viewpoint as well as by perturbative Feynman graphs, by many authors and can be found in a variety of works. [Fried and Gabellini (2009)] Because the renormalized QED coupling constant is so small, a few of the lowest orders of a perturbation expansion will usually suffice to produce a meaningful comparison with experiment; and the results have been so reassuring, that one must believe that QED is a true description of Nature. Aside from higher-order estimates in certain specific problems, the only open questions remaining in QED concern renormalization, and in particular, the logarithmic ultra-violet (UV) divergences which appear when perturbatively calculating the inverses of $Z_1 = Z_2$, and Z_3; here, $Z_{1,2,3}$ denote the vertex function renormalization constant, and the wave-function renormalization constants of the electron and photon, respectively.

Some years ago, it was suggested [Johnson, et al. (1967)] that perhaps sums of relevant Feynman graph contributions might show a cancellation of such UV divergences; but after a few unsuccessful attempts, that approach dissolved. One now understands that the possibility of such cancellation depends upon the summation of a truly infinite number of Feynman graphs involving closed-electron-loops each containing all possible, multiple photon exchanges across that loop, coupled by multiple photon exchanges of those "dressed" loops to all other relevant electron lines; and that the way to attack this problem, as for any "strong-coupling" problem is obviously via functional methods.

The thrust of the present QED presentation is to construct a functional representation of the radiative corrections to the photon propagator, leading to simple, proper-time representations for the second- and fourth-order terms. An intuitive, approximate extraction of the most important group of non-perturbative corrections needed for charge renormalization is performed; and with this summation over an infinite number of contributing graphs, one finds that the resulting, gauge-invariant, charge renormalization constant $[Z_3]^{(-1)}$ is finite, and it becomes clear why all previous perturbative and finite-kernel attempts have produced a divergent answer. Our evaluation is approximate, but physically and mathematically reasonable, and generates a renormalized $(\text{charge})^2/4\pi$ which is within one order of magnitude of $1/137$. Other questions are left untouched, such as the possible finiteness and gauge-invariance of mass renormalization, and the possible finiteness of the gauge-dependent renormalization constant Z_2^{-1}; and whether the techniques introduced here can be used, or modified, to determine if Z_2^{-1} is finite, at least within the set of relativistic gauges [Fried and Gabellini (2009)].[3]

We next consider in Chapter 7 the "fully dressed" electron propagator, and immediately see a possible limitation to using the previous photon analysis. However, another configuration-space approach suggests itself because the electron propagator is intrinsically gauge-dependent; and the choice of a special, simplifying gauge together with a renormalization-group approach generates an intuitive estimate of all radiative corrections in quenched approximation. The resulting wave-function-renormalization constant $[Z_2]^{(-1)}$ is still divergent, but use of the intuition behind the cancellation of the divergences of the dressed photon propagator is still possible; the difficulty here involves the interchange of two limiting procedures, and remains an open question.

[3]For a special choice of gauge, a summation over all Feynman graphs for the dressed electron propagator, in quenched approximation, has been performed in this paper, but it is not (yet) clear whether the resulting Z will be finite when all the closed-electron-loops are restored.

Chapter 8 defines a simple, heretofore overlooked QED model of "vacuum energy", one which has absolutely nothing to do with zero-point energies. This is a symmetry-breaking model, with a "bootstrap" solution, yielding a Lorentz-invariant vacuum energy which is everywhere finite. We emphasize that this model vacuum energy will not affect the motion of electrically-charged particles until the latter's energies are far greater than any obtainable from CERN's Large Hadron Collider. But the model is interesting for two reasons: Not only does it provide a QED-rooted explanation of Dark Energy and a possible connection with Inflation, but it suggests the existence of other, charged quantized fields bound together in the quantum vacuum; and these will be treated in detail in Part 4, to give a model basis for explaining some of the most puzzling astronomical observations.

Chapter 6

Radiative Corrections of the Photon Propagator

6.1 Functional Approach to the Photon Propagator

We begin with the GF of Chapter 3, extracted from (3.25) – (3.27). For the choice of the "free" photon propagator, we can adopt any of the "relativistic" gauges of form

$$D_{c,\mu\nu}(k) = \frac{1}{k^2 - i\epsilon}\left(\delta_{\mu\nu} - \rho\frac{k_\mu k_\nu}{k^2 - i\epsilon}\right) \tag{6.1}$$

in momentum space, or

$$D_{c,\mu\nu}(z) = \frac{i}{4\pi^2}\left(\frac{\delta_{\mu\nu}(1-\rho/2)}{z^2 + i\epsilon} + \rho\frac{z_\mu z_\nu}{(z^2 + i\epsilon)^2}\right) \tag{6.2}$$

in configuration space. Applying a pair of functional derivatives to the GF, and then setting all sources equal to zero, one builds the exact, or "dressed" photon propagator,

$$\begin{aligned} D'_{c,\mu\nu}(x-y) &= D_{c,\mu\nu}(x-y) + \int\int D_{c,\mu\lambda}(x-u)K_{\lambda\sigma}(u-w) \\ &\quad D_{c,\sigma\nu}(w-y)du\,dw \\ iK_{\mu\nu}(x-y) &= e^{\mathcal{D}_A}\frac{\delta}{\delta A_\mu(x)}\frac{\delta}{\delta A_\nu(y)}e^{L[A]}/ < S > |_{A=0} \\ &= e^{\mathcal{D}_A}\left[\frac{\delta^2 L}{\delta A_\mu(x)\delta A_\nu(y)} + \frac{\delta L}{\delta A_\mu(x)}\frac{\delta L}{\delta A_\nu(y)}\right]e^{L[A]}/ < S > |_{A=0} \end{aligned} \tag{6.3}$$

and the simplest, order-α term is just

$$iK^{(2)}_{\mu\nu}(x-y) = e^{\mathcal{D}_A}\frac{\delta^2 L}{\delta A_\mu(x)\delta A_\nu(y)}|_{A=0}. \tag{6.4}$$

It will be most relevant to remind the reader of certain rigorous properties of $L[A]$, which quantity in fact depends only upon $F_{\mu\nu}$, and can be easily written as such. This property immediately carries the consequence that currents induced in the vacuum, $< j_\mu(x) >= ig\delta L/\delta A_\mu(x)$, are to satisfy charge conservation: $\partial_\mu < j_\mu(x) >= 0$. In terms of the $K_{\mu\nu}$ of (6.3) and (6.4), this means that $\partial_\mu K_{\mu\nu} = \partial_\nu K_{\mu\nu} = 0$, so that the $\tilde{K}_{\mu\nu}(k)$ are expected to have the gauge invariant form $(k_\mu k_\nu - k^2\delta_{\mu\nu})\prod(k^2)$. The simplest, order $\alpha = g^2/4\pi$, Feynman graph corresponding to a single closed fermion loop does not display this property; and in the past, special, ad hoc maneuvers were invented to restore gauge invariance. In the Fradkin representation for $L[A]$ used in this book, gauge invariance to all orders is automatically satisfied.

We here sketch the evaluation of the lowest order $\tilde{K}^{(2)}_{\mu\nu}(k)$ directly from the gauge invariant $L[A]$ formalism, obtained from (6.4) without the linkage operations of that equation. The fermion determinant has an exact Fradkin representation:

$$L[A] = -\frac{1}{2}\int_0^\infty \frac{ds}{s} e^{-ism_0^2}\, e^{i\int_0^s ds' \frac{\delta^2}{\delta v_\mu^2(s')}}\, \delta^{(4)}\Big(\int_0^s ds' v(s')\Big)$$
$$\times \int d^4x' e^{-ig_0\int_0^s ds' v_\mu(s') A_\mu(x' - \int_0^{s'} v)} tr\Big(e^{g_0\int_0^s ds' \sigma_{\mu\nu}F_{\mu\nu}(x' - \int_0^{s'} v)}\Big)_+ \quad (6.5)$$

and we again find it useful to adopt the convenient reformulation of Section 5.3,

$$L[A] = -\frac{1}{2}\int_0^\infty \frac{ds}{s} e^{-ism_0^2}\int d^4x' N\int d[u] e^{\frac{1}{2}\int u(2h)^{-1}u}\delta^{(4)}(u(s))$$
$$\times e^{-ig_0\int_0^s ds' u'_\mu(s') A_\mu(x' - u(s'))} tr\Big(e^{g_0\int_0^s ds' \sigma_{\mu\nu}F_{\mu\nu}(x' - u(s'))}\Big)_+ \quad (6.6)$$

It is important to remember that there are two restrictions on the $u(s')$ variables, the first an implicit condition $u(s) = 0$, explicitly stated by the delta function of (6.6). For this lowest-order calculation, we replace m_0 by m.

Performing the pair of functional derivatives on $L[A]$ yields:

$$K^{(2)}_{\mu\nu}(x-y) = \frac{1}{2}\int_0^\infty \frac{ds}{s} e^{-ism^2} N\int d[u] e^{\frac{1}{2}\int u(2h)^{-1}u}\delta^{(4)}(u(s))$$
$$\times 4g_0^2\int_0^s ds_1\int_0^s ds_2\big[-u'_\mu(s_1)u'_\nu(s_2) - (\partial_\mu\partial_\nu - \delta_{\mu\nu}\partial^2)\big]$$
$$\delta(x - y + u(s_1) - u(s_2)) \quad (6.7)$$

after making use of the trace properties $tr[1] = 4$, and $tr[\sigma_{\alpha\mu}\sigma_{\beta\nu}] = \delta_{\alpha\nu}\delta_{\beta\mu} - \delta_{\alpha\beta}\delta_{\mu\nu}$. Writing a Fourier representation of $\delta(x-y+u(s_1)-u(s_2))$, the second line of (6.7) becomes:

$$\frac{4g_0^2}{(2\pi)^4}\int d^4k \int_0^s ds_1 \int_0^s ds_2\; e^{ik(x-y)+ik(u(s_1)-u(s_2))}$$
$$\times\big[-u'_\mu(s_1)u'_\nu(s_2) + (k_\mu k_\nu - \delta_{\mu\nu}k^2)\big] \tag{6.8}$$

and one can see that gauge invariance is maintained by imagining the result of calculating ∂_μ, or equivalently, of multiplying (6.8) by k_μ and summing over μ. The term coming from the sigma matrices obviously vanishes and the $u'(s_1)$ dependence yields:

$$\int_0^s ds_1 k\dot{u}'(s_1)e^{ik\dot{u}(s_1)} = -i\int_0^s ds_1 \frac{\partial}{\partial s_1}e^{ik\dot{u}(s_1)} = 0$$

because $u(0) = u(s) = 0$. At this early stage, it is then clear that the result of this computation must be gauge invariant.

In order to avoid any confusion on taking the $s_{1,2}$ derivatives of the u' factors with the $u(s_1) - u(s_2)$ exponential terms, it will be useful to introduce the sources $g_\mu(s') = k_\mu[\delta(s'-s_1) - \delta(s'-s_2)]$, and also $f\mu(s') = p_\mu\delta(s'-s)$, and there then follows:

$$\tilde{K}^{(2)}_{\mu\nu}(k) = 2ig_0^2\int_0^\infty \frac{ds}{s}e^{-ism^2}\int_0^s ds_1\int_0^s ds_2\int\frac{d^4p}{(2\pi)^2}N\int d[u]e^{\frac{i}{2}\int u(2h)^{-1}u}$$
$$\times\left[\frac{\partial}{\partial s_a}\frac{\partial}{\partial s_b}\frac{\delta}{\delta g_\mu(s_a)}\frac{\delta}{\delta g_\nu(s_b)} + (k_\mu k_\nu - \delta_{\mu\nu}k^2)\right]e^{i\int_0^s u\cdot(f+g)}\Bigg|_{\substack{s_a\to s_1\\ s_b\to s_2}}. \tag{6.9}$$

The normalized FI over the u dependence is then immediate, and yields the exponential factor:

$$e^{-i\int\int_0^s(f+g)\cdot h\cdot(f+g)} \tag{6.10}$$

so that the combinations $\frac{\partial}{\partial s_a}\frac{\partial}{\partial s_b}\frac{\delta}{\delta g_\mu(s_a)}\frac{\delta}{\delta g_\nu(s_b)}$ generate:

$$-4[p_\mu + k_\mu(\theta(s_1-s_a)-\theta(s_2-s_a))][p_\nu + k_\nu(\theta(s_1-s_b)-\theta(s_2-s_b))]$$
$$-\,2i\delta_{\mu\nu}\delta(s_a-s_b) \tag{6.11}$$

and the limit as $s_a \to s_1$, $s_b \to s_2$, produce, with $\theta(0) = \frac{1}{2}$,

$$(-2i)\delta_{\mu\nu}\delta(s_1-s_2)-4[p_\mu p_\nu+\frac{1}{4}k_\mu k_\nu+(p_\mu k_\nu+p_\nu k_\mu)(\theta(s_1-s_2)-\frac{1}{2})]. \quad (6.12)$$

With the exponential factor of (6.10) evaluated as:

$$-isp^2 - 2is_{12}p \cdot k - ik^2|s_{12}|, \qquad s_{12} = s_1 - s_2 \quad (6.13)$$

the $\int d^4p$ may be performed, leading to the cancellation of the term coming from the $tr[\sigma_{\alpha\mu}\sigma_{\beta\nu}$ dependent, and the result:

$$\left.\begin{aligned}\tilde{K}^{(2)}_{\mu\nu}(k) &= \tfrac{g_0^2}{8\pi^2}\textstyle\int_0^\infty \frac{ds}{s}e^{-ism^2}\frac{1}{s^2}\int_0^s ds_1\int_0^s ds_2\\ &\times\Big[2i\delta_{\mu\nu}\tfrac{1}{s} - 2i\delta_{\mu\nu}\delta(s_1-s_2)\\ &+4\tfrac{|s_1-s_2|}{s}(1-\tfrac{|s_1-s_2|}{s})k_\mu k_\nu - \delta_{\mu\nu}k^2\Big]\end{aligned}\right\} s_1 - s_2 = s_{12}$$

$$\times \exp\left[-isk^2\frac{|s_1-s_2|}{s}(1-\frac{|s_1-s_2|}{s})\right] \quad (6.14)$$

Finally, with the aid of the relations:

$$\int_0^s ds_1 \int_0^s ds_2 f(|s_1-s_2|) = 2s^2\int_0^1 dy(1-y)f(sy) \text{ with } y = \frac{|s_1-s_2|}{s}$$

$$\int_0^1 dy(1-2y)e^{-isk^2y(1-y)} = \frac{i}{sk^2}\int_0^1 dy\frac{\partial}{\partial y}e^{-isk^2y(1-y)} = o$$

and

$$\int_0^1 dy\, y(1-2y)e^{-isk^2y(1-y)} = \frac{i}{sk^2}\left[1-\int_0^1 dy\, e^{-isk^2y(1-y)}\right]$$

one obtains the result:

$$\tilde{K}^{(2)}_{\mu\nu}(k) = (k_\mu k_\nu - \delta_{\mu\nu}k^2)\Pi^{(2)}(k^2), \quad \Pi^{(2)}(k^2) = \int_0^\infty \frac{ds}{s}e^{-ism^2}\Pi^{(2)}(k^2,s) \quad (6.15)$$

in which the gauge symmetry has been preserved, and the subsequent $\int ds$ leads to a log divergence at its lower limit. With

$$\Pi^{(2)}(k^2,s) = \frac{g_0^2}{2\pi^2}\int_0^1 dy\, y(1-y)e^{-isk^2y(1-y)}$$

one has the simplest vacuum polarization result of a half-century ago [Schwinger (1949)]. Then, with the inverse of the photon's wave function renormalization constant given by

$$Z_3^{-1} = 1 + \Pi^{(2)}(0)$$

the renormalized, to order α, vacuum polarization is given by

$$\Pi^{(2)}(k^2) - \Pi^{(2)}(0)$$

which may easily be transformed into the more familiar form

$$\Pi^{(2)}(k^2) - \Pi^{(2)}(0) = -\frac{2\alpha}{\pi} \int_0^1 dy\, y(1-y) \ln\left(1 + y(1-y)\frac{k^2}{m^2}\right). \quad (6.16)$$

The real beauty of this calculation of this old and familiar result is that the UV divergence does not appear until the very last step, the integration over the proper time; and so does not disrupt the gauge symmetry of its elements. In contrast, the familiar Feynman graph computation in momentum space is so badly divergent that the underlying gauge symmetry is lost, and must be reinstated by other means. This was, of course, known to Schwinger, who originated proper time calculations in QFT; but it is made clear upon employing the elegant and most useful representations of Fradkin.

6.2 Higher-order Radiative Correction

In this section, we display the radiative corrections to the simplest, one-closed-fermion loop, and describe the cancellations which appear even before the corresponding FI is evaluated. These radiative corrections correspond to the action of the linkage operator upon the omitted A dependence of (6.3) and can be succinctly written by inserting the terms

$$e^{\mathcal{D}_a} e^{-ig_0 \int_0^s ds' u'_\mu(s') A_\mu(x'-u(s'))} tr\left(e^{g_0 \int_0^s ds' \sigma_{\mu\nu} F_{\mu\nu}(x'-u(s'))}\right)_+ \Big|_{A\to 0} \quad (6.17)$$

under all of the integrals of (6.6). Equations (6.10)–(6.13) are still relevant, but before the $\int d^4p$ and the $\int d[u]$ are performed, the u dependence of (6.17) must be extracted. For this, (6.17) may be rewritten, in the convenient form:

$$\left(e^{\mathcal{D}_A}e^{-ig_0\int_0^s ds' u'.A}\right)e^{\overleftrightarrow{\mathcal{D}}_A}\left(e^{\mathcal{D}_A}tr\left(e^{g_0\int_0^s ds'\sigma.F}\right)_+\right)\Big|_{A\to 0} \tag{6.18}$$

where $\overleftrightarrow{\mathcal{D}} = -i\int \frac{\overleftarrow{\delta}}{\delta A}D_c\frac{\overrightarrow{\delta}}{\delta A}$.

The first simplification to be noted was proven in Appendix B of [Fried and Gabellini (2009)],

$$e^{\mathcal{D}_A}tr\left(e^{g_0\int_0^s ds'\sigma\dot{F}}\right)_+\Big|_{A\to 0} = tr1 = 4$$

and can be trivially generalized to the more relevant statement:

$$e^{\mathcal{D}_A}tr\left(e^{g_0\int_0^s ds'\sigma.F}\right)_+ = tr\left(e^{g_0\int_0^s ds'\sigma.F}\right)_+$$

so that the self-linkages acting on this OE factor exactly cancel, to all orders in the coupling. In Feynman graph language, this would correspond to momentum space cancellations occurring in every higher order; in the Fradkin representation, one sees them immediately.

The self linkages $e^{\mathcal{D}_A}e^{-ig_0\int_0^s ds' u'.A}$ produce the dependence:

$$e^{i\frac{g_0^2}{2}\int_0^s ds_1\int_0^s ds_2\; u'_\mu(s_1)D_{c,\mu\nu}(u(s_1)-u(s_2))u'_\nu(s_2)}e^{-ig_0\int_0^s ds' u'\dot{A}} \tag{6.19}$$

leaving the cross linkage operation:

$$e^{-ig_0\int_0^s ds' u'.A}e^{\overleftrightarrow{\mathcal{D}}_A}tr\left(e^{g_0\int_0^s ds'\sigma.F}\right)_+\Big|_{A\to 0} \tag{6.20}$$

to be evaluated. The A dependence inside the OE can be extracted by writing the latter as:

$$\left(e^{2g_0\int_0^s \partial_\mu A_\nu\sigma_{\mu\nu}}\right)_+ = e^{-2ig_0\int_0^s ds'\partial_\mu A_\nu(x'-u(s'))\frac{\delta}{\delta\chi_{\mu\nu}(s')}}\left(e^{i\int_0^s ds''\sigma_{\mu\nu}\chi_{\mu\nu}(s'')}\right)_+$$

where $\chi_{\mu\nu}$ and $\delta/\delta\chi_{\mu\nu}$ are antisymmetric in μ and ν. In this way, the cross linkages produce the term:

$$e^{2ig_0^2\int_0^s ds_1\int_0^s ds_2 u'_\mu(s_1)\partial_\lambda D_{c,\mu\nu}(u(s_1)-u(s_2))}\frac{\delta}{\delta\chi_{\lambda_\nu}(s_2)} \tag{6.21}$$

which is then to act upon the waiting $\left(e^{i\int_0^s ds'\sigma_{\mu\nu}\chi_{\mu\nu}(s')}\right)_+$ factor and where $D_{c,\mu\nu} = \delta_{\mu\nu}D_c$, $D_c = \frac{i}{4\pi^2}[(\Delta\bar{u})^2 + i\epsilon]^{-1}$ and $\Delta u_\mu = u_\mu(s_1) - u_\mu(s_2)$; for simplicity, we are here using the Feynman gauge.

What is immediately clear is that the lowest order expansion of (6.21) will contribute nothing to the fourth order estimate of $\tilde{K}_{\mu\nu}(k)$, because a single $\delta/\delta\chi_{\lambda_\mu}$ operating upon $tr\left(e^{i\int\sigma\dot{\chi}}\right)_+\Big|_{\chi\to 0}$ produces the factor $tr[\sigma_{\lambda_\mu}] = 0$. $\tilde{K}^{(4)}_{\mu\nu}$ is therefore given precisely and exactly by the simple insertion of the g_0^2 order exponential term of (3.3) under the integrals defining $\tilde{K}^{(2)}_{\mu\nu}(k)$:

$$\tilde{K}^{(4)}_{\mu\nu}(k) = -g_0^4 \int_0^\infty \frac{ds}{s} e^{-ism^2} \int_0^s ds_1 \int_0^s ds_2 \int \frac{d^4p}{(2\pi)^4} N \int d[u] e^{\frac{i}{2}\int u(2h)^{-1}u}$$
$$\cdot \left[\frac{\partial}{\partial s_a}\frac{\partial}{\partial s_b}\frac{\delta}{\delta g_\mu(s_a)}\frac{\delta}{\delta g_\nu(s_b)} + (k_\mu k_\nu - \delta_{\mu\nu}k^2)\right] e^{i\int_0^s u\cdot(f+g)}$$
$$\int_0^s d\bar{s}_1 d\bar{s}_2 u'(s_1)\cdot u'(\bar{s}_2) D_c(u(\bar{s}_1) - u(\bar{s}_2)) \tag{6.22}$$

with D_c again in the Feynman gauge.

All that is now needed is to insert a Fourier representation for $D_c(u(\bar{s}_1) - u(\bar{s}_2))$:

$$\int \frac{d^4q}{(2\pi)^4} \frac{e^{iq\cdot(u(\bar{s}_1)-u(\bar{s}_2))}}{q^2 - i\epsilon} = i\int_0^\infty d\tau \int \frac{d^4q}{(2\pi)^4} e^{-i\tau q^2 + iq\cdot(u(\bar{s}_1)-u(\bar{s}_2))} \tag{6.23}$$

to replace the $\exp[iq\cdot(u(\bar{s}_1)-u(\bar{s}_2))]$ of (6.23) by $\exp[i\int_0^s j_\mu(s')u_\mu(s')]$ where $j_\mu(s') = q_\mu[\delta(s'-\bar{s}_1) - \delta(s'-\bar{s}_2)]$; to replace the $u'(\bar{s}_1)u'(\bar{s}_2)$ of (6.23) by $(\partial/\partial s_a)(\partial/\partial s_b)\left(\frac{1}{i}\right)^2(\delta/\delta g_\mu(s_a))(\delta/\delta g_\nu(s_b))$; and to then perform the gaussian integrals over u and q, before taking the limits $\bar{s}_{a,b} \to \bar{s}_{1,2}$. The result is a set of integrals over $\int_0^s ds_1 \int_0^s ds_2 \int_0^s d\bar{s}_1 \int_0^s d\bar{s}_2 \int_0^\infty dt$, to be followed by the final $\int_0^\infty \frac{ds}{s} e^{-ism^2}$; and this computation should be vastly simpler than that of the original Jost-Luttinger calculation [Jost and Luttinger (1950)].

There are two, related situations where linkages operating on OE terms may be expected to give a zero contribution to all orders, and these arise in the DP Model of Section 6.3, and somewhat differently in the Extended DP Model of Section6.4. In both of these cases we will be interested in a specific limit in which a Δu_μ variable approaches zero, and we will insist that these limits should be taken in a symmetric way, for both the odd function $\partial_\lambda D_c(\Delta u) = -\frac{i}{2\pi^2}\frac{\Delta u_\lambda}{[(\Delta u)^2+i\epsilon]^2}$, and the even function $\partial_\lambda\partial_\mu D_c(\Delta u) = -\frac{i}{2\pi^2}\frac{1}{[(\Delta u)^2+i\epsilon]^2}\left(\delta_{\mu\nu} - 4\frac{\Delta u_\mu \Delta u_\lambda}{[(\Delta u)^2+i\epsilon]}\right)$. In such a symmetric limit, as $\Delta u \to 0$, we will employ:

$$\partial_\lambda D_c(\Delta u)\big|_{\Delta u\to 0} = 0 \tag{6.24}$$

and

$$\partial_\lambda \partial_\mu D_c(\Delta u)\big|_{\Delta u\to 0} \sim \left(\delta_{\mu\nu} - \frac{4}{[(\Delta u)^2 + i\epsilon]} \cdot \frac{1}{4}\delta_{\mu\nu}(\Delta u)^2\right) = 0 \qquad (6.25)$$

and so replace (6.21) by unity, as well as the cross linkage operation between the OE terms of $\frac{\delta^2 L}{\delta A_\mu \delta A_\nu}$ and $\exp(L[A])$.

6.3 The DP Model for a Single Closed Fermion Loop

Return to the $\tilde{K}^{(2)}_{\mu\nu}(k)$ calculated in Section 6.1 written as:

$$\tilde{K}^{(2)}_{\mu\nu}(k) = (k_\mu k_\nu - \delta_{\mu\nu}k^2)\Pi^{(2)}(k^2), \quad \Pi^{(2)}(k^2) = \int_0^\infty \frac{ds}{s} e^{-ism^2}\Pi^{(2)}(k^2, s) \qquad (6.26)$$

where:

$$\Pi^{(2)}(k^2, s) = \frac{g_0^2}{2\pi^2}\int_0^1 dy\ y(1-y)e^{-isk^2 y(1-y)}.$$

We now insert the missing exponential part of this integrand, $\exp\left[i\frac{g_0^2}{2}\int u'\mathcal{D}_c u'\right]$ coming from all the remaining terms of $e^{\mathcal{D}_A}\frac{\delta^2 L}{\delta A \delta A}\Big|_{A=0}$ and retain only those parts which can contribute to subsequent UV divergences; these are the "dominant" parts, which define the DP Model, and in the following way.

As noted in Section 6.2, the choice of gauge is irrelevant, and we choose the simplest Feynman gauge, where the exponential factor $\frac{i}{2}\int u'\mathcal{D}_c u'$ becomes:

$$-\frac{g_0^2}{4\pi^2}\int_0^s ds_1 \int_0^s ds_2\ u'(s_1)\cdot u'(s_2)[(u(s_1) - u(s_2))^2 + i\epsilon]^{-1}. \qquad (6.27)$$

Since we are concerned in this and the following Section with a particular $\Delta u \to 0$ limit, all subsequent OE terms will be discarded.

It is intuitively clear that the most significant contributions will arise when $u(s_1)$ is close to $u(s_2)$, and we therefore expand $u(s_2)$ about (s_1), writing:

$$u_\mu(s_2) \simeq u_\mu(s_1) - (s_1 - s_2)u'_\mu(s_1) + \cdots$$

with $u(s_1)$ and $u'(s_1)$ considered as continuous functions. (The continuity of $u(s')$ is clear from its definition, as the integral over the Fradkin 4 velocity $v(s')$; while the continuity of $u'(s') = v(s')$ follows from the physical expectation that the 4 velocity of a particle, real or virtual, must be treated as a continuous function of its proper time parameter). All higher derivatives need not be continuous, and there is no obvious way of calculating them and their fluctuations; but they should not contribute to the leading divergent structures produced by the DP Model. This point is discussed and justified in detail in Appendix B.

To test the DP Model, in Appendix A we exhibit a completely independent, and simple perturbative example, whose log divergence is - to within additive constants - precisely the same as that calculated by the DP Model. But this point should be intuitively clear: because all u fluctuations are controlled by the gaussian weighting of the Fradkin representation, all u fluctuations must satisfy $u \leq \sqrt{s}$; and because all UV divergences arise from small s, and therefore from small $s_1 - s_2$, differences which scale as s, we may retain only the $(s_1 - s_2)u'(s_1)$ part of the denominator of (6.27), and replace the numerator $u'(s_2)$ by $u'(s_1)$. These simple replacements define the DP Model, and effectively permit the extraction of the leading divergence structure from under the FI of the u fluctuations.

The DP Model thus replaces (6.27) by the far simpler quantity:

$$-\frac{g_0^2}{4\pi^2}\int_0^s ds_1 \int_0^{s_1} ds' [u'(s_1)]^2 \left[s'^2 u'^2(s_1) + i\epsilon\right]^{-1}, \; s' = s_1 = s_2, \; u'^2 = u'_\mu u'_\mu$$

where it is understood that the effective UV cut off ϵ is to be held fixed until the very last step of all calculations. If $[u'(s_1)]^2 = 0$, this integral vanishes; if not, with $\alpha_0 = g_0^2/4\pi$ it can be rewritten as:

$$-\frac{\alpha_0}{\pi}\int_0^s ds_1 \int_0^{s_1} ds' \left[s'^2 + i\epsilon\bar{\epsilon}(s_1)\right]^{-1}$$

where we denote by $\bar{\epsilon}(s_1)$ the sign of $[u'(s)]^2$. And since $\int_0^{s_1} ds' \left[s'^2 + i\epsilon\bar{\epsilon}(s_1)\right]^{-1}$ can be rewritten as $\int_0^{s_1} ds' \left[s' + i\epsilon\bar{\epsilon}(s_1)\right]^{-2} = -\frac{1}{s_1 + i\epsilon\bar{\epsilon}} + \frac{1}{i\epsilon\bar{\epsilon}}$, the first of the two needed integrations can be trivially performed.

We now insist that the second term immediately above, $\frac{1}{i\epsilon}\int_0^s ds_1 \frac{1}{\bar{\epsilon}(s_1)} = \frac{1}{i\epsilon}\int_0^s ds_1 \bar{\epsilon}(s_1)$, must vanish, and for the following several reasons. (a) Intuitively, nothing in the formalism distinguishes between positive and negative values of $[u']^2$, and therefore the sum over the sign of all possible u'^2 fluctuations must vanish. (b) If this integral did not vanish, there would appear in all orders of perturbation theory its coefficient $1/\epsilon$, which corresponds to a quadratic UV divergence; and in all known examples of perturbative, gauge invariant, QED calculations, such quadratic UV divergences are absent. (c) One can point to related, if indirect, arguments, such as treating the FI of the Fradkin representation as a normalized probability function, and using it to calculate the value of the expected sign of the integral of u'^2, which also turns out to be zero. Henceforth, we shall assume this most reasonable property, that the integral over the sign of u'^2 vanishes.

With this understanding, our desired exponential factor is modeled by:

$$\frac{\alpha_0}{\pi}\int_0^s ds_1[s_1 + i\epsilon\bar{\epsilon}(s_1)]^{-1}.$$

Since we expect $\bar{\epsilon}$ to fluctuate, as the integral over s_1 proceeds, and since the knowledge of $\bar{\epsilon}$ is only necessary near the lower limit of the s integral, we can rewrite the integral as:

$$\frac{\alpha_0}{\pi}\int_\epsilon^s \frac{ds_1}{s_1} = \frac{\alpha_0}{\pi}\ln\left(\frac{s}{\epsilon}\right) \tag{6.28}$$

which quantity misses a possible $\frac{\alpha_0}{\pi}\ln(i\bar{\epsilon}(0))$ additional term, depending on the sign of $\bar{\epsilon}(s_1 \to 0)$. For the moment, we suppress this factor, retaining the obvious $\ln(s/\epsilon)$ dependence.

The exponential of (6.28) is our DP Model of all the radiative corrections (of this Section) to $\prod^{(2)}(k^2)$, and yields for the sum of those contributions:

$$\Pi^{(2)}(k^2) = 2\frac{\alpha_0}{\pi}\int_0^1 dy\; y(1-y)\int_\epsilon^\infty \frac{ds}{s}\left(\frac{s}{\epsilon}\right)^{\alpha_0/\pi} e^{-is[m_0^2+k^2y(1-y)]} \tag{6.29}$$

where we have cut off the lower limit of the s integral at ϵ, and will later use the identification $\epsilon \to 1/\Lambda^2$. Although (6.29) contains log divergent terms in every order, it is interesting to note that those additional radiative corrections conspire to remove the necessity of using an ϵ as the lower limit of the s integral if $\alpha_0 > 0$. In any finite order perturbation calculation, that lower limit ϵ cut off is absolutely necessary; but since we have summed over an infinite number of graphs, there has occurred a qualitative change in the s integrand, such that it was not necessary to introduce this lower limit (although the need for a cut off is simply transferred to large values of s). There are still divergences, in every order (of an α_0 expansion), but this effect suggests the possibility that the omitted radiative corrections, arising from the linkage of all possible closed fermion loops to those loops under consideration above, may tend to suppress the divergences which remain in (6.29). And this is indeed the case.

Restoring the neglected phase factor of (6.28), we calculate $Z_3^{-1} = 1 + \prod(0)$ by rotating the contour of the s integration to run along the negative imaginary axis, and so obtain:

$$\Pi(0) = \frac{\alpha_0}{3\pi} \int_{\epsilon m_0^2}^{\infty} \frac{d\tau}{\tau} \left(\frac{\tau}{m_0^2 \epsilon}\right)^{\alpha_0/\pi} \exp\left[\frac{\alpha_0}{\pi} \ln(-i) - \frac{\alpha_0}{\pi} \ln(i\bar{\epsilon}(0))\right] e^{-\tau m^2} \tag{6.30}$$

and note that the reality of Z_3 fixes the choice $\bar{\epsilon}(0) = -1$.

Changing variables, in (6.30) to $\tau = \epsilon m_0^2 x$ produces:

$$\Pi(0) = \frac{\alpha_0}{3\pi} \int_1^{\infty} dx\ x^{\alpha_0/\pi - 1} e^{-\epsilon m_0^2 x} \simeq \frac{\alpha_0}{3\pi} \int_1^{1/\epsilon m_0^2} dx\ x^{\alpha_0/\pi - 1}$$

or

$$\Pi(0) \simeq \frac{1}{3}\left[\left(\frac{\Lambda^2}{m_0^2}\right)^{\frac{\alpha_0}{\pi}} - 1\right] \simeq \frac{1}{3}\left(\frac{\Lambda^2}{m_0^2}\right)^{\frac{\alpha_0}{\pi}} \tag{6.31}$$

as long as $\alpha_0 >$ and $(\Lambda/m_0) \gg 1$. Z_3^{-1} can now be represented by (6.31); or by an infinite sequence of log divergent terms:

$$Z_3^{-1} = 1 + \frac{1}{3}\sum_{n=1}^{\infty} \frac{1}{n!}\left(\frac{\alpha_0}{\pi}\right)^n \ln^n\left(\frac{\Lambda^2}{m_0^2}\right). \tag{6.32}$$

An alternate representation of Z_3^{-1} is obtained by allowing the lower limit of the integral of (6.30) to approach 0, which yields:

$$\Pi(0) = \frac{\alpha_0}{3\pi}\left(\frac{\Lambda^2}{m_0^2}\right)^{\frac{\alpha_0}{\pi}} \int_0^{\infty} d\tau e^{-\tau} \tau^{\alpha_0/\pi - 1} = \frac{1}{3}\left(\frac{\Lambda^2}{m_0^2}\right)^{\frac{\alpha_0}{\pi}} \Gamma(1 + \frac{\alpha_0}{\pi})$$

and then:

$$Z_3^{-1} = 1 + \frac{1}{3}\left(\frac{\Lambda^2}{m_0^2}\right)^{\frac{\alpha_0}{\pi}} \Gamma(1 + \frac{\alpha_0}{\pi}). \tag{6.33}$$

The gauge invariant part of the photon propagator, in this approximation, is given by:

$$\tilde{D}_c'(k)(\delta_{\mu\nu} - k_\mu k_\nu / k^2)$$

where:

$$[\tilde{D}_c'(k)]^{-1} = k^2[1+\Pi(k^2)] \tag{6.34}$$

and $\prod(k^2)$ is the sum over those proper self energy terms (which cannot be constructed from an iteration over lower order terms) calculated in DP approximation. The simplest renormalization procedure in QED proceeds by adding and subtracting $\prod(0)$ in the inverse of the denominator of (6.34), so that:

$$\tilde{D}_c'(k) = (k^2)^{-1}[1+\Pi(0)+\Pi(k^2)-\Pi(0)]^{-1}$$

and the wave function renormalization constant Z_3 identified as the coefficient of the k^2 pole of $\tilde{D}_c'(k)$ as $k^2 \to 0$, which leads to the familiar identification of Z_3^{-1} in terms of $1+\prod(0)$. The renormalized propagator is:

$$\tilde{D}_{c',R}(k) = Z_3^{-1}\tilde{D}_c'(k)$$

or:

$$\tilde{D}_{c',R}(k) = (k^2)^{-1}[1+Z_3(\Pi(k^2)-\Pi(0))]^{-1} \tag{6.35}$$

where the combination $Z_3(\prod(k^2)-\prod(0))$ defines, in each sequential order, a finite contribution given in terms of the α constructed from the Z_3 contribution of that order, and α_0 is chosen to have whatever (large) value is required so that the renormalized $\alpha = 1/137$.

In our DP Model, the calculation of $\prod(k^2)-\prod(0)$ is again immediate, yielding:

$$\begin{aligned}\Pi(k^2)-\Pi(0) = -\frac{1}{3}\left(\frac{\Lambda^2}{m_0^2}\right)^{\frac{\alpha_0}{\pi}}\Gamma(1+\frac{\alpha_0}{\pi})\\ \left[1-6\int_0^1 dy\; y(1-y)\left(1+\frac{k^2}{m_0^2}y(1-y)\right)\right]^{-\frac{\alpha_0}{\pi}}\end{aligned} \tag{6.36}$$

and:

$$Z_3[\Pi(k^2)-\Pi(0)] = -\left[1-6\int_0^1 dy\; y(1-y)\left(1+\frac{k^2}{m^2}y(1-y)\right)\right]^{-\frac{\alpha_0}{\pi}}. \tag{6.37}$$

The fact that our DP result for large (k/m) does not agree in form with the Jost–Luttinger perturbative result is irrelevant, for there is no reason to expect the DP model, which deals with the extraction of log divergences, to provide the correct limiting values of perturbative quantities, such as the coefficients of the leading, and finite, $\alpha^2 \ln^2(k/m)$ dependence, although such a possibility may be approximately realized. While our DP Model sums up all the possible perturbative divergences to the gauge invariant photon renormalization, it misses possibly important additive k/m dependence to those divergent logs, for it is computing in the region of $k = 0$, rather than large k/m. But the DP method does show, as in Section (6.4), for the $k = 0$ computation of Z_3^{-1}, that it is sensible to consider this gauge invariant sector of QED as a finite QFT; and this, conceptually, is a new and most satisfying result.

Finally, to understand what it is that our DP Model achieves, we ask and answer the following question. How can one understand the connection between the vanishing of a propagator's denominator in this configuration space, functional formulation and the appearance of log divergences in momentum space? Simply by taking the Fourier transform of $D_c(u(s_1)-u(s_2))$ and asking why does the corresponding

$$\int \frac{d^4k}{(2\pi)^4} \frac{e^{ik\cdot(u(s_1)-u(s_2))}}{k^2}$$

converge when combined with another momentum-space propagator? It is not the insertion of a denominator factor proportional to $[(q-k)^2 + M^2]^{-1}$ coming from another propagator, bosonic or fermionic, for that alone generates a log divergence. Rather, it is the $\exp[ik \cdot (u(s_1) - u(s_2))]$ term - which of course is eventually evaluated in and by the $\int d[u]$ FI - which provides enough oscillations and cancellations to yield a finite result. But when $u(s_1)-u(s_2)$ vanishes, as we have suggested above, that log divergence will appear and will enter all relevant parts of the computations of that order. We find it far simpler to remain in functional configuration space, where the cancellations of the OE terms are easily visible, then to convert to the conventional Feynman graph analysis in momentum space. The divergent logs will reappear, but the cancellations obvious in functional configuration space will there require much tedious calculation to obtain. It is far more efficient to identify and extract in their functional source those log divergences, than to perform the FIs, convert to momentum space, distribute those divergences in the conventional parts of a Feynman graph, and then attempt, order by order, to understand and remove them.

6.4 The Extended DP Model, and the Finiteness of Z_3

We return to (6.3) and the exact $K_{\mu\nu}$, from which follows the complete Z_3^{-1}. The first task to perform is to argue that the:

$$e^{\mathcal{D}_A} \frac{\delta L}{\delta A_\mu(x)} \frac{\delta L}{\delta A_\nu(y)} e^{L[A]} / < S > \Big|_{A=0}$$

terms cannot contribute to Z_3^{-1}, and for this we return to the Fradkin representation for $L[A]$, here simplified by the neglect of all QE terms, as discussed in the previous Section. Together with the Proof of Appendix B, this means that all OE dependence of every $L[A]$ does not contribute to the radiative corrections calculated in the DP Models.

Consider first the product $\frac{\delta L}{\delta A_\mu(x)} \frac{\delta L}{\delta A_\nu(y)}$, which contains under its separate integrals the terms relevant to this discussion:

$$\int d^4x' \delta(u(s)) \int_0^s ds_1 u'_\mu(s_1) \delta(x' - u(s_1) - x) \int d^4x'' \delta(\bar{u}(\bar{s}))$$
$$\int_0^{\bar{s}} d\bar{s}_1 \bar{u}_\nu(\bar{s}_1) \delta(x'' - \bar{u}(s_1) - y) \times \exp\big[- ig_0$$
$$\int_0^s ds' u'_\alpha(s') A_\alpha(x' - u(s')) - ig_0 \int_0^{\bar{s}} d\bar{s}' \bar{u}'_\beta(\bar{s}') A_\beta(x'' - \bar{u}(\bar{s}'))\big].$$

Neglecting for the moment cross linkages to the $\exp(L[A])$ term, we then have:

$$e^{\mathcal{D}_A} \frac{\delta L}{\delta A_\mu(x)} e^{\overleftrightarrow{\mathcal{D}}_A} e^{\mathcal{D}_A} \frac{\delta L}{\delta A_\nu(y)} \Bigg|_{A=0} =$$

$$\int d^4x' \delta(u(s)) \int d^4x'' \delta(\bar{u}(\bar{s})) \int_0^s ds_1 \int_0^{\bar{s}} d\bar{s}_1 u'_\mu(s_1) \bar{u}'_\nu(\bar{s}_1) \delta(x' - u(s_1) - x)$$
$$\delta(x'' - \bar{u}(\bar{s}_1) - y) e^{i\frac{g_0^2}{2} \int \int_0^s ds'_1 ds'_2 D_c(u(s'_1) - u(s'_2)) u'(s'_1) \cdot u'(s'_2)}$$
$$e^{i\frac{g_0^2}{2} \int \int_0^s d\bar{s}'_1 d\bar{s}'_2 D_c(\bar{u}(\bar{s}'_1) - \bar{u}(\bar{s}'_2)) \bar{u}'(\bar{s}'_1) \cdot \bar{u}'(\bar{s}'_2)}$$
$$e^{ig_0^2 \int_0^s d\hat{s}_1 \int_0^{\bar{s}} d\hat{s}_2 u'(\hat{s}_1) \cdot \bar{u}'(\hat{s}_2) D_c(x' - x'' + \bar{u}(\hat{s}_2) - u(\hat{s}_1))}. \tag{6.38}$$

With the aid of the delta functions of (6.38), its last line may be rewritten as:

$$e^{ig_0^2 \int_0^s d\hat{s}_1 \int_0^{\bar{s}} d\hat{s}_2 u'(\hat{s}_1) \cdot \bar{u}'(\hat{s}_2) D_c(x - y + \Delta(u, \bar{u}))} \tag{6.39}$$

and the $\int d^4x' \int d^4x''$ performed so that (6.38) reduces to:

$$\delta(u(s)) \delta(\bar{u}(\bar{s})) \int_0^s ds_1 u'_\mu(s_1) \int_0^{\bar{s}} d\bar{s}_1 \bar{u}'_\nu(\bar{s}_1) \mathcal{S}(\Delta u) \mathcal{S}(\Delta \bar{u})$$
$$e^{ig_0^2 \int_0^s d\hat{s}_1 \int_0^{\bar{s}} d\hat{s}_2 u'(\hat{s}_1) \cdot \bar{u}'(\hat{s}_2) D_c(x - y + \bar{u}(\hat{s}_2) - u(\hat{s}_1))} \tag{6.40}$$

where $\Delta(u, \bar{u}) = u(s_1) - u(\hat{s}_2) - \bar{u}(\bar{s}_1) + \bar{u}(\hat{s}_2)$, and $\mathcal{S}(\Delta u)$ and $\mathcal{S}(\Delta \bar{u})$ refer to the self linkage exponentials of (6.38).

The renormalized charge in QED is conventionally defined by evaluating $\tilde{K}_{\mu\nu}(k)$ at $k = 0$, corresponding to the definition of physically measured charge at large distances, specifically at distances large compared to the Compton wavelength m^{-1}; the radiative corrections occur at distances less than m^{-1}, and in the context of these calculations, this corresponds to evaluating $\tilde{K}_{\mu\nu}(x-y)$ at separations $x - y \gg m^{-1}$. But all of the $x - y$ dependence of (6.40) lies in the D_c corresponding to its cross-linked exponential factor, while the u and $\bar{u}$ quantities, from their definitions scale as $\sqrt{s} \sim \sqrt{\bar{s}} \sim m^{-1}$. It is then clear that in the limit of $x - y \gg m^{-1}$, the Δu and $\Delta \bar{u}$ dependence of $D_c(x - y + \Delta(u, \bar{u}))$ is effectively suppressed, and in this limit the factors $\int_0^s d\hat{s}_1 u'(\hat{s}_1)$ and $\int_0^{\bar{s}} d\hat{s}_2 \bar{u}'(\hat{s}_2) = 0$, so that this exponential factor completely disappears. In a similar way, after the x' and x'' integrations have been performed, the entire contribution of (6.40) is itself proportional to similar factors, $\int_0^s ds_1 u'_\mu(s_1) \int_0^{\bar{s}} d\bar{s}_1 \bar{u}_\nu(\bar{s}_1)$ which also vanish.

Insertion of the cross linkages between (6.38) and the A dependence of $\exp(L[A])$ does not change this situation, for the limit of large $x - y$ in the cross linked terms can be understood as the separate limits of $x \to \infty$ and $y \to -\infty$, so that these cross linkages also vanish. The computation then reduces to that of (6.38) as the self-linkages of $\exp(L[A])$ are cancelled by the definition of $< S >$. The result is that the entire quantity:

$$e^{\mathcal{D}_A} \frac{\delta L}{\delta A_\mu(x)} \frac{\delta L}{\delta A_\nu(y)} e^{L[A]} / < S > \Bigg|_{A=0}$$

does not contribute to Z_3.

The first term of (6.3) is the relevant quantity, and we first consider the structure of the factor $e^{\mathcal{D}_A} e^{L[A]}$. It will be convenient to express the latter in terms of the functional cluster expansion:

$$e^{\mathcal{D}_A} e^{L[A]} = \exp\left[\sum_{n=1}^{\infty} \frac{1}{n!} Q_n[A]\right]$$

where $Q_n[A] = e^{\mathcal{D}_A} (L[A])^n \Big|_{\text{connected}}$, that is: $Q_1[A] = e^{\mathcal{D}_A} L[A] = \bar{L}[A]$, $Q_2[A] = (e^{\mathcal{D}_A} L)(e^{\overleftrightarrow{\mathcal{D}}_A} - 1)(e^{\mathcal{D}_A} L)$, etc. The reason for choosing this expansion is that the $Q_n, n > 1$, can be estimated to yield smaller values than does Q_1 (as well as being far more difficult to calculate); their divergence structures are similar to that of Q_1, but, as discussed in Appendix C, they play a smaller role int he overall calculation. Q_1 alone is sufficient to remove all the perturbative log divergences of Z_3^{-1}.

Rather than repeat all the details of every equation in the next few paragraphs, we shall simply present the added features that arise from the cross linkages between the $e^{\mathcal{D}_A}\frac{\delta^2 L}{\delta A_\mu(x)\delta A_\nu(y)}$ of the previous Section and the $e^{\mathcal{D}_A} e^{L} \simeq e^{\bar{L}}$ of the present discussion. Note that all the self linkages of $e^{\mathcal{D}_A} e^{L[A]}/ < S > |_0$ simply disappear from the final result by virtue of the definition of $< S >$. We are therefore interested in:

$$\left(e^{\mathcal{D}_A}\frac{\delta^2 L}{\delta A_\mu(x)\delta A_\nu(y)}\right) e^{\overleftrightarrow{\mathcal{D}}_A} e^{\bar{L}[A]}\Big|_{A\to 0} \tag{6.41}$$

which, continuing to use the variables $u_\mu(s')$ for the FI of $e^{\mathcal{D}_A}\frac{\delta^2 L}{\delta A_\mu(x)\delta A_\nu(y)}$ corresponds to the insertion under the latter's FI the quantity:

$$\exp\left[-g_0\int_0^s ds' u'_\alpha(s')\int d^4w D_c(x'-u(s')-w)\frac{\delta}{\delta A_\alpha(w)}\right] e^{\bar{L}[A]}\Big|_{A\to 0} \tag{6.42}$$

where we again hold to the Feynman gauge.

But the operation of (6.42) is just a translation operator, and regardless of what it acts upon, has the effect of shifting the A dependence of that function — in this case $\bar{L}[A]$ — by the quantity:

$$\exp\left[ig_0^2\int_0^s ds' u'_\alpha(s')\int_0^t dt' v'_\alpha(t') D_c(x'-x''-u(s')+v(t'))\right] \tag{6.43}$$

appearing under the FI of $\bar{L}[A]$, with x'' and $v_\alpha(t')$ the variables of that functional. In addition to the factor of (6.43), there appear under the $\bar{L}$ FI the self linkages of amount:

$$\exp\left[i\frac{g_0^2}{2}\int\int_0^t dt_1 dt_2 D_c(v(t_1)-v(t_2))v'(t_1)\cdot v'(t_2)\right]$$

which are independent of x' and x''. In effect, what the cross linkages have achieved is to insert s dependence under and mixed with the t integrals of $\bar{L}$; and subsequent integration of that t dependence generates an exponential s dependence which will have a damping effect on all the divergent expressions of Section 6.5. This simple observation is at the heart of the mechanism for obtaining a finite charge renormalization.

The self linkages of $\bar{L}$ may be read off from those calculated in Section 6.3, Equation (6.28) and its subsequent discussion:

$$\exp\left[\, i\frac{g_0^2}{2}\int\int_0^t dt_1 dt_2 D_c(v(t_1)-v(t_2))v'(t_1)\cdot v'(t_2)\right] \to \left(\frac{t}{i\epsilon, \cdot\bar{\epsilon}(0)}\right)^p,$$

$$p = \alpha_o/\pi.$$

The cross linkage term of (6.43) can be evaluated in a similar manner, with the realization that only the continuous parts of the functions $u(s')$ and $v(t')$ are relevant — as noted in Appendix B — and in essence they are very similar functions, differing mainly in their physical place in the calculation. They represent the continuous parts of the fluctuations defined by the same functionals; and it is reasonable to ask when they can interact and combine directly with each other, for when $u(s') \approx v(t')$ one has the beginning of an incipient divergence. A modification of the previous DP model is now defined by expanding $v(t') \simeq v(s') + (t'-s')v'(s') + \cdots$, for the case when these functions are essentially the same: when $u(s') \simeq v(t')$ and $u'(s') \simeq v'(t')$, so that the denominator of the cross linkages becomes:

$$\left[((x'-x'') + (t'-s')v'(s'))^2 + i\epsilon\right] \tag{6.44}$$

or the same form with v replaced by u. Since the functional integrations sum over all possible (continuous) forms, which fluctuations are defined in exactly the same way, there should be strong possibility of such overlaps. The essence of this Extended DP Model is that $u(s')$ and $v(t')$ are treated in a completely symmetric manner; and this requirement of symmetry turns out to be a guarantee of simplicity of the forms that follow. This intuitive assumption defines the Extended DP Model, in which the difference of two, equivalent, continuous fluctuations of identical functionals has the same possibility of overlap as in the DP Model. We emphasize that we cannot prove the validity of this assumption but it is most certainly intuitive; and it forms the basis of the cancellations of divergent logarithms that are about to occur.

However, even the EDP Model cannot guarantee the vanishing of (6.44), for another difference, $x'-x''$, need not be small, and in most cases is not. What is the consequence when $x'-x''$ is large? Quite independently of the EDP Model, when $x'-x'' \gg m^{-1}$, the Compton wavelength of the charged fermion traveling about the loop, that difference completely dominates $u(s')-v(t')$, because each of the latter quantities scale as $\sqrt{s} \sim \sqrt{t} \sim m^{-1}$, and hence the $u-v$ difference in that case is irrelevant. But then, as repeatedly emphasized, the s' and t' integrals of (5.6) vanish,

$\int_0^s ds' u'_\alpha(s') \int_0^t dt' v'_\alpha(t') = 0$ and remove the entire cross-linkage term from consideration. When does this not happen? Only when $x' - x''$ is restricted to values on the order of, or less than $u(s') - v(t')$. But we are interested in small differences, where s' and t' tend to the order of ϵ, and where subsequently $\epsilon \to 0$. How can this be arranged?

We shall here assume that the only contribution to the cross-linkage integral comes when $|x' - x''| \simeq \xi\sqrt{\epsilon}$, where $\xi \sim O(1)$. SInce this is an idealization, one must expect fluctuations about this condition, such that ξ will turn out to be somewhat less than unity. For conceptual simplicity, choose the point x' as the origin of the x'' coordinates, and consider a (Euclidean) 4-sphere of radius $\xi\sqrt{\epsilon}$. For any point within this sphere, $|x' - x''|$ effectively disappear from (5.7), and we can apply the EDP Model; this means that the only non-zero values of the x'' integrals is given by:

$$\int d^4x'' \to \int d\Omega_4 \int_0^{\xi\sqrt{\epsilon}} r^3 dr = \frac{\pi^2}{2}\xi^4\epsilon^2. \tag{6.45}$$

Can (6.45) produce a non-zero result? Yes, because $\bar{L}$ is itself proportional to the factors $-\frac{1}{2}tr[1] \int d^4x'' \int_\epsilon^\infty \frac{dt}{t} e^{-itm^2} N \int d[v] \exp\left[\frac{i}{2}\int v(2h)^{-1}v\right]\delta^{(4)}(v(t))$ and, just as for the DP Model, the evaluation of $N\int d[v]$ yields: $N\int d[v] \exp\left[\frac{i}{2}\int v(2h)^{-1}v\right]\delta^{(4)}(v(t)) = -i\frac{\pi^2}{t^2}\frac{1}{(2\pi)^4}$, so that the entire set of exponential integrals reduces to:

$$T\left(\frac{s}{\epsilon}\right) = i\left(\frac{\xi}{2}\right)^4 \epsilon^2 \int_\epsilon^\infty \frac{dt}{t^3} e^{-itm_0^2} e^{3i\pi p/2} \left(\frac{s}{\epsilon}\right)^p \left(\frac{t}{\epsilon}\right)^{2p}. \tag{6.46}$$

In (6.46), one factor of $e^{i\pi p/2}\left(\frac{t}{\epsilon}\right)^p$ arises from the self linkages of $\bar{L}$, while the cross linkages generate the remaining factors. We have maintained strict s, t symmetry by writing:

$$\begin{aligned}
&\int_0^s ds' \int_0^t dt' \frac{u'(s')\cdot v'(t')}{(u(s') - v(t'))^2 + i\epsilon} \\
&= \int_0^s ds' \int_0^t dt' \left[\frac{\theta(s'-t')}{(s'-t'+i\epsilon\bar{\epsilon}(s'))^2} + \frac{\theta(t'-s')}{(t'-s'+i\epsilon\bar{\epsilon}(t'))^2}\right] \\
&= \int_0^s ds' \left[\frac{1}{i\epsilon\bar{\epsilon}(s')} - \frac{1}{s'+i\epsilon\bar{\epsilon}(s')}\right] + \int_0^t dt' \left[\frac{1}{i\epsilon\bar{\epsilon}(t')} - \frac{1}{(t'+i\epsilon\bar{\epsilon}(t'))}\right].
\end{aligned}$$

Again, the integrals over the $\bar{\epsilon}$ factors vanish, and the result is simply:

$$-\ln\left(\frac{s}{i\epsilon\bar{\epsilon}(0)}\right) - \ln\left(\frac{t}{i\epsilon\bar{\epsilon}(0)}\right)$$

which is properly symmetric in s and t. Were that symmetry not preserved, the results would lead to far more complicated forms, requiring detailed numerical integrations in order to verify the expectations of a finite Z_3^{-1} and an α close to 1/317. In contrast, the symmetric Extended DP Model adopted here leads to results obtainable in closed form, and to the immediate verification of our expectations.

With the inclusion of these cross linkages, the DP integral representation for Z_3^{-1} is changed to:

$$Z_3^{-1} = 1 + \left(\frac{p}{3}\right) e^{i\pi p/2} \int_\epsilon^\infty \frac{ds}{s} e^{-ism_0^2} \left(\frac{s}{\epsilon}\right)^p e^{T(s/\epsilon)}. \tag{6.47}$$

Note that we are keeping to the conventional perturbative form (although in configuration, rather than momentum space) of cutting off all proper time integrals with a lower limit of ϵ, which will shortly be set equal to zero. Without the cross-linkage factors that produce T, (6.47) has the same divergences as docs (6.30); but an entirely new situation now arises with the insertion of (6.46) into (6.47). Proper time contours need not be rotated; all that is needed is the simple change of variables: $s/\epsilon = x$, $t/\epsilon = y$, so that (6.47) may be rewritten as:

$$Z_3^{-1} = 1 + (\tfrac{p}{3}) e^{i\pi p/2} \int_1^\infty dx \; x^{p-1} e^{-ixm_0^2\epsilon} e^{T(x)} \tag{6.48}$$

where:

$$T(x) = i\left(\frac{\xi}{2}\right)^4 e^{3i\pi p/2} x^p \int_1^\infty \frac{dy}{y^3} e^{-i\epsilon y m_0^2} y^{2p}$$

and one notes that, for convergence as $y \to \infty$, one must have $p < 1$. For $p > 0$, $T(x)$ can act as a damping or oscillating factor that provides convergence for the x integral as $x \to \infty$; and for both the x and y integrals, we may now sagely let $\epsilon \to 0$, and the divergences have disappeared.

As this program is carried out, one notes the independence of Z_3 on m_0, or on m. This property is not at all clear from perturbation theory, where sequential renormalization of mass and charge, along with simultaneous changes in $Z_{1,2}$ must appear. In fact, Z_3, and therefore, α, are independent of the charged particle's mass, in agreement with the experimental fact that all charged fermions obeying QED (but not simultaneously QCD) have the same electric charge.

What remains is to insure that Z_3^{-1} is real, and hence the condition $\mathrm{Im}Z_3^{-1} = 0$ specifies a relation that $p = \alpha_0/\pi$ must satisfy. In principle, this is true; and if that condition leads to a single allowed value of α_0, and hence of α, one will have solved an old and deep question in Physics. In our calculation, however, there appears the parameter ξ, as a measure of the difficulty and uncertainty of extracting the divergent character of (6.44). And with the neglect of higher terms of the cluster expansion, there is no guarantee that a single value of α_0 will emerge. Rather, with the cluster approximation already made, and with those approximations we are about to make, we are gratified to find a range of p values, $0 < p < 1$, within which we can choose α_0 such that $\alpha \sim 10^{-2}$.

For $p < 1$, the integral defining $T(x)$ is readily obtained, and the expression for Z_3^{-1} now reads:

$$Z_3^{-1} = 1 + \left(\frac{p}{3}\right) e^{i\pi p/2} \int_1^\infty dx\ x^{p-1} e^{-qx^p} \tag{6.49}$$

where the q parameter is given by:

$$q = -i \left(\frac{\xi}{2}\right)^4 \frac{e^{3i\pi p/2}}{2(1-p)}.$$

Note that convergence of the x integral is obtained for $Re(q) > 0$, which corresponds to $p < 2/3$, and that a necessary but not sufficient condition for the removal of $\mathrm{Im}Z_3$ is that $p > 1/3$.

The x integral is completely trivial because its integrand is a perfect differential, and one finally obtains:

$$Z_3^{-1} = 1 + \frac{2}{3}(1-p)\left(\frac{\xi}{2}\right)^{-4} ie^{-i\pi p} e^{-q}. \tag{6.50}$$

Since we expect ξ to be somewhat less than 1, the ration $(\xi/2)^4$ should be very small, and a good approximation to (6.50) is then:

$$Z_3^{-1} = 1 + \frac{2}{3}(1-p)\left(\frac{\xi}{2}\right)^{-4} \big(\sin(\pi p) + i\cos(\pi p)\big) \tag{6.51}$$

and it is clear that $p = 1/2$ insures that the Z_3^{-1} of (6.51) is real, while the choice $\xi \simeq 0.397$ lead to an $\alpha = \pi p Z_3$ of approximately $1/137$. From this solution, one calculates $(\xi/2)^4 \simeq 0.00155$, so that any correction to these parameters obtained by the use of (6.50) rather than (6.51) cannot differ from the above values of p and ξ by more than a few parts per thousand.

6.5 Summary

The thrust of this presentation has been to argue, by summing the "naturally divergent" terms of all relevant radiative corrections, that charge renormalization in QED is finite. We do not claim to have given a mathematically rigorous proof of that statement, but rather an intuitive statement, based on the functional structure of QED. We have argued that our extraction of logarithmically divergent terms corresponds to those found in lower order radiative corrections using Feynman graph techniques; and we believe that momentum space, graphical techniques become impossible, ad therefore irrelevant, in any attempt to include all, or almost all, radiative corrections of arbitrarily high order.

The functional techniques we use are based upon a convenient rearrangement of the Schwinger/Symanzik functional solution for the generating functional of QED, together with a slight rearrangement of Fradkin's most useful functional representation for Green's functions $G_c[A]$ and, in particular, for $L[A]$, the log of the so called "fermion determinant". $L[A]$ contains the basic, gauge invariance of the photon propagator, and that structure is here realized by means of most convenient linkage operations.

If a perturbative expansion is desired, this functional approach will exactly reproduce the conventional Feynman graphs, but it has the great advantage of working in configuration rather than momentum space, and one can take advantage of cancellations which occur there before any computation is required, but which are achieved in momentum space only after painful and tedious manipulations. [Fried and Gabellini (2009)] and, it should be noted, that frequently, as is the case for the lowest-order radiative corrections to the photon propagator, singularities of the Feynman integrals can mask the symmetry structure of the theory, an unpleasant attribute of that method of calculation which is quite absent from functional methods built around proper time representations[1].

Within this formalism, and the intuition we have used to obtain the results described above, we have derived a pair of integrals in (6.48), with ϵ set equal to zero, which should express the finite character of charge renormalization. And we have illustrated a possible solution of those equations with a simplified model which generates a specific value of α_0 that leads to an $\alpha \sim 1/137$. What this model solution does <u>not</u> do is to obtain a

---

[1]Functional representations provide a simple realization of the "Last Rule", often stated and rarely understood, for writing the totality of all Feynman graphs of a given order: "Sum over all topologically distinct graphs".

single, precise and necessary value of α_0, and so determine α. As noted above, because of the imprecision in extracting the divergences from the cross linked, closed fermion loop functionals, as expressed by the parameter ξ, this goal may be elusive; and we will have to be content with choosing an α_0 which does reproduce the experimental α. Previous attempts have been made some decades ago [Johnson, et al. (1967)][Baker and Johnson (1969)][Gell-Mann and Low (1954)] to simultaneously display a cancellation of divergences, which might lead to a value of α close to its renormalized value. These were noble efforts, especially in the context of Feynman graphs; and one can now see why they were unsuccessful, for the crucial aspect of including an infinite number of closed fermion loops, each containing all possible photonic "dressing" in a manifestly gauge invariant way, could not be done. It is gauge invariance, built into the Fradkin representation for $L[A]$, along with the use of proper time techniques which preserve that invariance, which allows one to identify and extract divergences.

It should be noted also that the assumptions made in these papers, which center about a single log divergence for Z_3^{-1} persisting in higher perturbative orders, together with the special choice of a zero bare fermion mass, are quite different from what we observe. Our Z_3^{-1} is finite and independent of mass. The use of a finite number of terms in a Bethe-Salpeter kernel (cf [Johnson, et al. (1967)], Section 1) misses an infinite class of Feynman graphs, for a new class of graphs appears in each higher order; and in each missing class there are an infinite number of Feynman graphs for higher and higher orders. In contrast, we make no assumptions about the nature of perturbative divergences in Z_3^{-1}; we include them as we see them, and we are able to see and include them because we use a formulation where their appearance is obvious, and where strict gauge invariance helps us to extract them, and sum them to all orders, literally.

These are other, related questions to which one does not (yet) know the answers, which now present themselves, demanding attention. Is it possible that, within a class of gauges, Z_1 and Z_2 are also finite? Mass renormalization should be a completely gauge-invariant affair; and is it finite? Can these techniques be extended to QCD, to show that color charge renormalization is finite? Are Faddev-Popov ghosts really necessary in a truly gauge invariant formulation of QCD? From our experience in QED, it is clear that manifest gauge invariance, or the protection of a desired symmetry, automatically, in every order of approximation of a Schwinger/Symanzik/Fradkin functional context, is the essential ingredient.

Perhaps the best way to end this Section is by emphasizing that while the functional tools used are surely powerful, and appropriate for this problem, the identification and extraction of divergences has been intuitive. Nevertheless, the obvious advantages of this functional approach to the calculation of radiative corrections seems clear, and deserve strong emphasis.

Chapter 7

Radiative Corrections to the Electron Propagator

7.1 Introduction

This Chapter is a somewhat amplified version of a recent publication [Fried and Gabellini (2009)], describing a somewhat different method of approach to the simplest of QED fermonic amplitudes, the "dressed" electron propagator, containing almost all of its radiative corrections (in quenched approximation). Although within the functional formulation emphasized in this book, this treatment will be heuristic, containing apparently-reasonable approximations which - and this should be stressed - lack a proper mathematical justification. The interested reader is urged to attempt to prove, or disprove, those few limiting assertions which yield a pleasantly compact, final representation for the quenched $S_c'(x-y)$.

A few lines near the end of the previous Chapter considered the application of the techniques used for the dressed photon propagator, and in particular the question of removing the quenched approximation in this treatment of the electron propagator. The same question was raised in a recent publication [Fried, et al. (2012)], and there it was concluded that the techniques used for the estimation of Z_3 are inappropriate for a similar attempt at Z_2. The analysis of this Chapter will come to the same conclusion but in a different way; and that difference is worth a few words, for it is suggested by the fact that S_c' is itself a gauge-dependent quantity, and the question then arises: Might there be a special choice of gauge within which the quenched calculation simplifies immensely?

With the aid of an intuitive mathematical limit of a simple re-scaling operation, the answer to that question is positive; and it leads to a redefinition of the problem in terms of a soluble equation of Renormalization Group form. This solution is given in configuration space, and so does not immediately contain a non-perturbative evaluation of mass renormalization, whose possible finiteness and gauge invariance remain open questions. Rather, the simple course of action followed in these pages is to produce a solution for the quenched electron propagator in configuration space, to then neglect the difference between bare and renormalized masses, and then easily obtain the corresponding Z_2. In the context of a non-perturbative approach, perhaps the most relevant aspect of this solution is its multiplicative character, as a simple product of a free propagator multiplying an appropriate function of its configuration space variable containing all powers of the square of the coupling constant. One may expect this multiplicative aspect to appear as part of the complete, non-perturbative solution for S'_c; but that remains an open question.

7.2 Formulation

We begin by stating the functional form of the dressed electron propagator, discussed in Chapter 5, and immediately move to its quenched approximation,

$$S'_c(x-y) = e^{\mathcal{D}_A} G_c(x,y|A) \frac{e^{L[A]}}{< S >}\Bigg|_{A\to 0} \to e^{\mathcal{D}_A} G_c(x,y|A)\Bigg|_{A\to 0}, \quad (7.1)$$

where $< S >= e^{\mathcal{D}_A}\, e^{L[A]}\big|_{A\to 0}$, $\mathcal{D}_A = -\frac{i}{2}\int \frac{\delta}{\delta A_\mu} D_c^{\mu\nu} \frac{\delta}{\delta A_\nu}$ and $G_c[A] = [m_0 + \gamma\cdot(\partial - igA)]^{-1}$.

One then introduces the exact Fradkin representation of (5.39),

$$\begin{aligned} G_c(x,y|A) &= C^{-1} e^{-\frac{1}{2}Tr\ln(2h)} \cdot i\int_0^\infty ds\, e^{-ism_0^2} \cdot \int d[u] = e^{\frac{i}{2}\int u(2h)^{-1}u} \\ &\cdot [m_0 - \gamma\cdot(\partial - igA(x))] \cdot \delta^{(4)}(x-y+u(s)) \\ &\cdot e^{-ig\int_0^s ds' u'(s')\cdot A(y-u(s'))} \cdot \left(e^{g\int_0^s ds' \sigma\cdot F(y-u(s'))}\right)_+ \end{aligned} \quad (7.2)$$

and makes explicit the A-dependence inside its OE, using the argument following (6.19). The linkage operator $\exp\{\mathcal{D}_A\}$ is then to be applied (from the left) upon all the terms of (7.2).

When the linkage operator is passed through the multiplicative $\gamma\cdot A(x)$ factor, one obtains, with $z = x - y$,

$$e^{\mathcal{D}_A}\gamma \cdot A(x)\cdots\Big|_{A\to 0} \to -i\gamma_\mu \int d^4x' D_c(x-x')\frac{\delta}{\delta A_\mu(x')}\cdots\Bigg|_{A\to 0}$$

which will, in turn, "bring down" from the exponential factors of (7.2) terms proportional to

$$\int_0^s ds'\ u'(s')D_c(z-u(s'))\ldots$$

and to

$$\gamma_\alpha\gamma_\beta \int_0^s ds'\ \partial_\beta^z D_c(z-u(s'))\cdots .$$

But these factors are expected to vanish under the scaling operation to be performed below, and in the interests of simplicity and clarity they will be suppressed at this point.

It should be emphasized that it is also at this point that this presentation loses the ability to discuss mass renormalization, because under the scaling transformation the $u(s')$ are to increase in magnitude, while the z_μ are held fixed. But going to the mass shell, where mass renormalization is defined, corresponds to z^2 becoming very large; and it is precisely this interchange of limits which cannot be trusted. With this understanding, we bypass the proper analysis of mass renormalization, and simply replace the bare mass, m_0 by the renormalized mass, m.

With $S_c'(z) = (m - \gamma\cdot\partial^z)\mathcal{S}_c'(z)$, the dressed, quenched, electron propagator can then be written in the form

$$\begin{aligned} \mathcal{S}_c'(z) = i\int_0^\infty ds\ e^{-ism^2} \cdot e^{-\frac{1}{2}Tr\ln(2ah)} \cdot N' \int d[u]\ e^{\frac{i}{2}\int u(2ah)^{-1}u} \\ \cdot\delta^{(4)}(z+u(s)) \cdot e^{\frac{ig^2}{2}\int\int_0^s ds_1 ds_2 u_\mu'(s_1)\ D_c^{\mu\nu}(\zeta)u_\nu'(s_2)} \\ \cdot e^{-2ig^2\int\int_0^s ds_1 ds_2 \frac{\delta}{\delta X_{\mu\lambda}(s_1)}\overrightarrow{\partial}_\lambda D_c^{\mu\nu}(\zeta)\overleftarrow{\partial}_\gamma \frac{\delta}{\delta X_{\nu\gamma}(s_2)}} \\ \cdot e^{-g^2\int\int_0^s ds_1 ds_2 u_\mu'(s_1) D_c^{\mu\nu}(\zeta)\overleftarrow{\partial}_\lambda \frac{\delta}{\delta X_{\nu\lambda}(s_2)}} \cdot \left(e^{\int_0^s ds'\sigma_{\alpha\beta}X_{\alpha\beta}(s')}\right)_+\Bigg|_{X\to 0}, \end{aligned} \tag{7.3}$$

where $\zeta_u = u_\mu(s_1) - u_\mu(s_2) \equiv \Delta u_\mu(s_1,s_2)$, $h(s_1,s_2) = \frac{1}{2}(s_1+s_2-|s_1-s_2|)$, N' is a normalization constant depending on the Δs partitions of the FI, and a is a real, positive number to be set equal to 1 at the end of the calculation.

As in Section 6.2, one notes the functional simplification — corresponding to a cancelation amongst higher-order Feynman graphs — that to all g^z orders,

$$e^{-2ig^2 \int \int_0^s ds_1 ds_2 \frac{\delta}{\delta X} (\vec{\partial} D_c \overleftarrow{\partial}) \frac{\delta}{\delta X}} \cdot \left(e^{\int_0^s \sigma \cdot X} \right)_+ \Big|_{X \to 0} = 1. \tag{7.4}$$

However, the $u' \ldots$ OE do not appear to vanish in a similar way, but generate log divergent terms in every higher g^2 order.

A second and most useful observation may now be made, concerning the form of the relativistic gauge-dependent, free-photon propagator used in (7.1), which, in momentum space takes the form

$$\tilde{D}_{c,\mu\nu}(k) = \frac{1}{k^2 - i\epsilon} \left(\delta_{\mu\nu} - \rho \frac{k_\mu k_\nu}{k^2 - i\epsilon} \right),$$

but in configuration space becomes

$$D_c^{\mu\nu}(z) = \frac{i}{4\pi^2} \left(\frac{\delta_{\mu\nu}[1 - \rho/2]}{z^2 + i\epsilon} + \rho \frac{z_\mu z_\nu}{(z^2 + i\epsilon)^2} \right).$$

Conventional choices of the parameter ρ, with $\rho = 0, 1, -2$, define the Feynman, Landau, and Yennie gauges, respectively.

A special gauge is here defined by the choice $\rho = +2$, for in this gauge the $u' \ldots u'$ term of (7.3) may be rewritten in the form

$$\begin{aligned} &\exp\left[-\frac{g^z}{4\pi^2} \int \int_0^s ds_1 ds_2 \frac{u'_\mu(s_1) \Delta u_\mu}{(\Delta u)^2 + i\epsilon} \cdot \frac{u'_\nu(s_2) \Delta u_\nu}{(\Delta u)^2 + i\epsilon} \right] \\ &= \exp\left[\gamma \int \int_0^s ds_1 ds_2 \frac{\partial}{\partial s_1} \ln Z \cdot \frac{\partial}{\partial s_2} \ln Z \right], \end{aligned} \tag{7.5}$$

with $Z = M^2[(\Delta u)^2 + i\epsilon]$, $\gamma = g^2/4\pi'$, and M an arbitrary (for the moment) mass parameter introduced for dimensional reasons. Equation (7.5) is almost a perfect (double) differential, but not quite; but it can be rewritten as

$$\exp\left[\gamma \int \int_0^s ds_1 ds_2 \left(\frac{1}{2} \frac{\partial}{\partial s_1} \cdot \frac{\partial}{\partial s_2} \ln^2 Z - \ln Z \cdot \frac{\partial}{\partial s_1} \cdot \frac{\partial}{\partial s_2} \ln Z \right) \right] \tag{7.6}$$

in which the first term of (7.6) is a perfect differential, and as such its u-fluctuations of the FI all cancel away, and its value is given by the end-point $u(s) = -z,\ u(0) = 0$ quantities as

$$\exp\left[\frac{\gamma}{2}\int_0^s\int \frac{\partial^2}{\partial s_1 \partial s_2}\ln^2 Z\right]$$
$$= \exp\left[-\gamma\ln^2\left(\frac{z^2+i\epsilon}{i\epsilon}\right) - 2\gamma\ln\left(\frac{z^2+i\epsilon}{i\epsilon}\right)\cdot\ln(i\epsilon M^2)\right]. \tag{7.7}$$

Evaluation of the second term of (7.6) was first attempted by approximation. Since the ϵ of $\ln Z$ acts as a cut-off parameter in configuration space (in momentum space, $\epsilon \sim \Lambda^{-2}$), one expects that $\ln Z$ should be "slowly varying", and the second term of (7.6) could reasonably be approximated by

$$\exp\left[-\gamma\langle\ln Z\rangle\right]\int\int_0^s ds_1 ds_2 \frac{\partial^2}{\partial s_1 \partial s_2}\ln Z] \tag{7.8}$$

where the $\langle$ average $\rangle$ is taken over $s_{1,2}$ and the u-fluctuations. The integrals of (7.8) are now perfect differentials, which can be evaluated immediately:

$$\int\int_0^s ds_1 ds_2 \frac{\partial^2}{\partial s_1 \partial s_2}\ln Z = -2\ln\left(\frac{z^2+i\epsilon}{i\epsilon}\right), \tag{7.9}$$

and the remaining functional integration is trivial, yielding the free-particle result, $I_o = (4\pi^2 as)^{-2}\cdot e^{iz^2/4as}$. This procedure was originally thought to be the beginning of a strong-coupling approximation; but one can do much better, as follows.

7.3 Computation

Add and subtract to the second term of (7.6) the quantity

$$-\gamma\ln Q\int\int_0^s ds_1 ds_2 \frac{\partial^2}{\partial s_1 \partial s_2}\ln Z \tag{7.10}$$

where Q is a real, positive number > 1. In the remaining FI,

$$R(z^2,a) = e^{-\frac{1}{2}Tr\ln(2ah)} \cdot N' \int d[u] e^{\frac{i}{2}\int\int u(2ah)^{-1}u} \delta^{(4)}(z+u(s))$$
$$\cdot \exp\left[-\gamma \int\int ds_1 ds_2 \ln Z \frac{\partial^2}{\partial s_1 \partial s_2} \ln Z\right]$$
$$\cdot e^{-g^2 \int\int_0^s ds_1 ds_2 u'_\mu(s_1) D_c^{\mu\nu}(\zeta) \overleftarrow{\partial}_\lambda \frac{\delta}{\delta X_{\nu\lambda}(s_2)}}$$
$$\cdot \left(e^{\int_0^s ds' \sigma_{\alpha\beta} X_{\alpha\beta}(s')}\right)_+ \Big|_{X\to 0}, \tag{7.11}$$

this replaces the term $\exp[-\gamma \int\int ds_1, ds_2 \ln Z \cdot \frac{\partial^2}{\partial s_1 \partial s_2} \ln Z]$ by

$$\exp\left[-\gamma \ln Q \int\int ds_1 ds_2 \frac{\partial^2}{\partial s_1 \partial s_2} \ln Z - \gamma \right.$$
$$\left. \int\int ds_1 ds_2 \ln(Z/Q) \frac{\partial^2}{\partial s_1 \partial s_2} \ln(Z/Q)\right]. \tag{7.12}$$

But $\ln(Z/Q)$ can be written as $\ln\left(M^2 \frac{(\Delta u)^2 + i\epsilon}{Q}\right) = \ln\left(M^2\left[\frac{(\Delta u)^2}{Q} + i\epsilon\right]\right)$, and rescaling the dummy variables $u_\mu(s_i) = \sqrt{Q}\bar{u}_\mu(s_i)$ consistently in (7.11) produces the statement:

$$R(z^2,a) = \frac{1}{Q^2} e^{2\gamma \ln Q \cdot \ln\left(\frac{z^2+i\epsilon}{i\epsilon}\right)} \cdot I\left(\frac{z^2}{Q}, \frac{a}{Q}, \frac{1}{Q}\right), \tag{7.13}$$

where

$$I\left(\frac{z^2}{Q}, \frac{a}{Q}, \frac{1}{Q}\right) = e^{-\frac{1}{2}Tr\ln(2\frac{a}{Q}h)} \cdot N' \int d[\bar{u}] e^{\frac{i}{2}\int\int \bar{u}(2\frac{a}{Q}h)^{-1}\bar{u}} \cdot \delta^{(4)}\left(\bar{u}(s) + \frac{z}{\sqrt{Q}}\right)$$
$$\cdot e^{-\gamma\int\int \ln Z(\Delta\bar{u}) \frac{\partial^2}{\partial s_1 \partial s_2} \ln Z(\Delta\bar{u})}$$
$$\cdot e^{\frac{-g^2}{Q}\int\int \bar{u}'_\mu D_c^{\mu\lambda}(\Delta\bar{u}) \overleftarrow{\partial}_\nu \frac{\partial}{\partial X_{0\lambda}}} \cdot \left(e^{\int \sigma\cdot X}\right)_+ \Big|_{X\to 0}. \tag{7.14}$$

Because of the $1/Q$ dependence of the exponential OE term, (7.13) is not a useful scaling relation. But if Q is taken as arbitrarily large - far larger than any logarithmically divergent, perturbatively-calculated $u' \ldots$ OE terms - then all of those terms are individually removed, and their sum appears in the exponential factor

$$\exp\left[2\gamma \ln Q \cdot \ln\left(\frac{z^2+i\epsilon}{i\epsilon}\right)\right].$$

If this procedure is mathematically correct, it then represents a magnificent calculational tool, for now the OE dependence — which has always blocked non-perturbative estimations (except in Bloch–Nordsieck approximations, where such terms are neglected) — has been effectively summed.

Without going into the details of whether this limit is true, which the author would prefer to leave to a more mathematically-capable reader, the validity of this limit will here be accepted as a reasonable conjecture. For now, with $R(z^2, a)$ independent of Q, one has a useful scaling relation,

$$R(z^2, a) = \frac{1}{Q^2} e^{2\gamma \ln Q \cdot \ln\left(\frac{z^2 + i\epsilon}{i\epsilon}\right)} \cdot I\left(\frac{z^2}{Q}, \frac{a}{Q}\right), \tag{7.15}$$

where $I\left(\frac{z^2}{Q}, \frac{a}{Q}\right)$ denotes the FI (7.14) without the $u' \ldots$ OE term. It should be noted that the OE terms are themselves gauge-invariant, since they stem from an initial $F_{\mu\nu}$ dependence; but this special gauge provides the framework for their summation.

Renormalization Group methods can now be employed to provide a differential equation for $I(\frac{z^2}{Q}, \frac{a}{Q})$, by considering a small variation of the very large Q, and calculating $0 = Q\frac{\partial}{\partial Q} R(z^2, a)$,

$$0 = \left[2\gamma \ln\left(\frac{z^2 + i\epsilon}{i\epsilon}\right) - 2\right] I\left(\frac{z^2}{Q}, \frac{a}{Q}\right) - \left(z\frac{\partial}{\partial z^2} + a\frac{\partial}{\partial a}\right) I\left(\frac{z^2}{Q}, \frac{a}{Q}\right). \tag{7.16}$$

It is simplest to rescale $z^2 \to z^2 \cdot Q$, $a \to a \cdot Q$ to obtain the differential equation

$$\left\{\left(z^2\frac{\partial}{\partial z^2} + a\frac{\partial}{\partial a}\right) + 2\left[1 - \gamma \ln Q - \gamma \ln\left(\frac{z^2 + i\epsilon}{i\epsilon}\right)\right]\right\} I(z^2, a) = 0. \tag{7.17}$$

We look for a solution of form $I(z^2, a) = I_0(z^2, a) J(z^2, a)$, where $I_0(z^2, a) = S_c'|_{g\to 0}$, because out of the "landscape" of possible solutions to the functional Schwinger equations, this is the one desired. Further, one can demand that $J(z^2, a)\big|_{a\to 1, z^2\to 0} = 1$, so that the equal-time anti-commutation relations originally assumed for free and dressed fermion operators remain the same. With $J = \exp\Omega$, (7.17) becomes

$$\left(z^2\frac{\partial}{\partial z^2} + a\frac{\partial}{\partial a}\right)\Omega(z^2, a) = 2\gamma\left[\ln Q + \ln\left(\frac{z^2 + i\epsilon}{i\epsilon}\right)\right], \tag{7.18}$$

and this is the relation that one must now solve.

Inspection shows that to any solution of (7.18) satisfying the above boundary conditions can be added an arbitrary function of z^2/a which respects those conditions. This lack of mathematical uniqueness, however, is not detrimental to our physical description because the only task of the dressed electron propagator in any scattering or production amplitude is to produce a wave-function renormalization constant, Z_2, which can be identified, as needed, in every n-point function, so as to cancel with an equal Z_1, and thereby provide a gauge-independent renormalization of the electron's charge. Z_2 is gauge dependent, and therefore not a measurable quantity.

Whichever solution is chosen, the essence of this special gauge + rescaling method is that the non-perturbative solutions for this dressed propagator are essentially multiplicative in coordinate space. In contrast, Feynman graphs in higher orders of perturbation theory involve horrendous and overlapping integrals in momentum space. This special gauge is the bridge that leads to analytically-obtainable solutions in coordinate space. Of course, Fourier transforms must finally be taken; but that is a separate matter.

At the end of this calculation, the parameter $a \to 1$. Since there is no a-dependence on the RHS of (7.18), one may take as the simplest solution that obtained by assuming $\Omega = \Omega(y), y = \ln[M^2(z^2 + i\epsilon)]$, and (7.18) then becomes $\frac{d\Omega}{dy} = 2\gamma[y + \ln\left(\frac{Q}{i\epsilon M^2}\right)]$, with solution $\Omega(y) = \gamma y^2 + 2\gamma y \ln(Q/i\epsilon M^2) + k$, with k constant, so that:

$$J(z^2, a) = \exp\left[\gamma \ln^2\left(M^2(z^2 + i\epsilon)\right) + 2\gamma \ln\left(M^2(z^2 + i\epsilon)\right)\ln(Q/i\epsilon M^2) + k\right]. \tag{7.19}$$

Remembering to rescale$z^2 \to z^2/Q$ (in order to undo the passage from (7.16)to(7.17)),and including the previous, perfect differential term of(7.7),

$$\frac{\gamma}{2}\int\int_0^s ds_1 ds_2 \frac{\partial^2}{\partial s_1 \partial s_2} \ln^2 Z = -\gamma \ln^2\left(\frac{z^2 + i\epsilon}{i\epsilon}\right) - 2\gamma \ln\left(\frac{z^2 + i\epsilon}{i\epsilon}\right)\ln(i\epsilon M^2)$$

the entire answer becomes:

$$S_c'(z) = I_0(z^2, 1)\exp\left[2\gamma \ln\left(\frac{z^2 + i\epsilon}{i\epsilon}\right)\ln\left(\frac{Q}{i\epsilon M^2}\right)\right.$$
$$\left. + 2\gamma \ln Q \ln(i\epsilon M^2) - \gamma \ln^2(i\epsilon M^2) + k\right].$$

If the exponential factor multiplying I_0 is to become unity when $z^2 \to 0$, then the constant of this solution must be chosen as $k = \gamma \ln^2(i\epsilon M^2) - 2\gamma \ln Q \ln(i\epsilon M^2) - is\delta m^2$, where the term $-is\delta m^2$ has been included in this constant so as to renormalize the bare mass sitting in the exponential factor of (7.2). The entire result, with m_0 everywhere replaced by m is then:

$$S_0'(z) = I_0(z^2, 1) \exp\left[2\gamma \ln\left(\frac{z^2 + i\epsilon}{i\epsilon}\right) \ln\left(\frac{Q}{i\epsilon M^2}\right)\right] \tag{7.20}$$

where the result of all interactions is, in configuration space, simply a multiplicative, log divergent, exponential factor multiplying the free field propagator.s

7.4 Summary

Without actually performing the Fourier transform into momentum space let us try to guess the electron's wave function renormalization (WFR) constant. Going to the mass shell in momentum space corresponds to taking the limit $z^2 \to \infty$ in coordinate spae. This can be represented as an infra red limit in momentum space, if z^2 is represented by $1/\mu^2$; also, as noted above, $\epsilon \sim \Lambda^{-2}$ can be thought of as an UV cut off in momentum space. The argument of the multiplicative exponential factor of (7.20) then becomes:

$$2\gamma\left[-\left(\frac{\pi}{2}\right)^2 + \ln\left(\frac{\Lambda^2}{\mu^2}\right)\ln\left(\frac{Q\Lambda^2}{M^2}\right)\right] - i\pi\gamma\ln\left(\frac{\Lambda^4 Q}{\mu^2 M^2}\right)$$

and in order for the Z_2 of this solution to be real, one must choose $\left(\frac{\Lambda^4 Q}{\mu^2 M^2}\right) = 1$, or $M^2 = Q\frac{\Lambda^4}{\mu^2}$, so that:

$$Z_2 = \exp\left[2\gamma\left(\left(\frac{\pi}{2}\right)^2 + \ln^2\left(\frac{\Lambda^2}{\mu^2}\right)\right)\right] \tag{7.21}$$

is not only real, but is bounded between 0 and 1, as expected from the formal theory. Calculation of the Fourier transform into momentum space, and the identification of Z_2 as the coefficient of the renormalized mass-shell pole, leads to the same answer as that of (7.21) above, and is performed in Appendix C of [Fried and Gabellini (2009)].

Finally, we return to the question concerning the possibility of employing the methods of Chapter 6 to extract the non-quenched electron propagator. Unfortunately, this does not seem to be immediately possible because the conditions stated in that Chapter for the validity of the EDP Model are no longer satisfied. Instead of the properties $u(s) = u(0) = v(t) = v(0)$ of the photon calculation, which allowed one to expect the similarity of the functions $u(s)$ and $v(t)$, for the electron propagator one has $u(0) = v(t) = v(0) = 0$, but $u(s) = z$, the variable conjugate to the electron's momentum. And as one goes to the mass shell in p^2, one expects z^2 to become very large, so that there could be a considerable difference in the (continuous) variations of $u(s)$ and $v(t)$. For this reason, the intuition of the EDP Model tends to disappear, and there remains no obvious mechanism to cancel the divergencies, as was done by the function $T(x)$ for Z_3. On the basis of this argument, we see no alternative to the conclusion expressed long ago by [Kallen (1953)], who conjectured that at least one of the renormalization constants of QED is divergent. From this perspective, it is the gauge-dependent, unmeasurable Z_2 which diverges, while the gauge-independent Z_3 is finite.

Chapter 8

A QED Symmetry-Breaking Model of Vacuum Energy

8.1 Introduction

Every so often, one is struck by an idea which is so simple and compelling that it is hard to understand why it has been overlooked. In our QFT considerations of the very small, such an idea appears with astrophysical implications on a grand scale. The essence of this very simple idea is as follows. Conventionally, electromagnetic fields are either "quantized" or "external", and by external is meant classical fields which can be switched on and off. Simultaneously, one speaks of the "quantum vacuum", in which quantized fields of arbitrary complexity are fluctuating, and whose effects can only be indirectly inferred (such as the 27 Megacycles of the 1057 Megacycle Lamb Shift, separating the $2S^{1/2}$ and $2P^{1/2}$ levels of the Hydrogen atom.

Imagine that one had available a "super Heisenberg" gamma-ray microscope, so that one could "see" a virtual photon of the QM vacuum suddenly transform itself into a bubble corresponding to the virtual appearance of an electron and positron, whose propagations define the sides of the bubble, and which propagation continues for a duration on the order of the inverse of their mass. When these virtual particles are separated there is an electric field between them, which can be thought of an electromagnetic fluctuation that disappears when the bubble collapses. Such fluctuations are present in the QM vacuum; they cannot be "turned off") and, in an appropriately averaged way, they should contain energy. They arise from the fundamental quantization of the operator QED fields. The averaged or background electromagnetic energy of these fluctuations might well be characterized by the existence of an effective, c-number $A_{\mu}^{\rm vac}$ which is always present. The existence of such an averaged, effective field, clue to its QM origin, could possibly have striking effects on the classical world about us.

The implications of this idea suggest a sensible theory of Inflation, as the Precursor of Dark Energy, as well as an obvious extension to a Theory of Dark Matter, and thence to a possible way of understanding observations of Gamma-ray emissions of extraordinary energy — perhaps even including the recent "Fermi Bubbles" — along with a mechanism for Ultra-high-energy Cosmic rays that are able to violate the GZK limit. One is even lead to a possible glimpse of the Creation of our Universe. These topics will be discussed in detail, along with relevant theoretical and observational references, in Part 4 of this Book; in this Chapter, the basic model for a QED Vacuum Energy is derived and applied to a description of Dark Energy. A previous suggestion for this effect [Fried (2003)] described a different symmetry-breaking solution, but one applicable only to Dark Energy; the present, "improved" solution can provide a simple, physical picture for the onset, duration, and magnitudes of Inflation, as well as present-day Dark Energy.

At this point it may be useful to consider just what is here meant by Vacuum Energy, especially in order to understand what is meant by "effective Lorentz invariance". In the context of our Model, for clarity now simplified to include only lowest-order loop fluctuations, one imagines charged fermion loops fluctuating in and out of existence, with the normal vector to the plane of their loop fluctuating in all possible directions. At the instant that any given loop appears there is (thinking classically) an electric field across that loop, between the oppositely charged fermions forming the sides of the loop; and that field signifies that at that instant there is a certain amount of electrostatic energy present, which apparently disappears when the loop collapses. But then another loop, with its normal vector pointing in a different direction appears; and then another, and another, such that it is clear that, on average, there is a certain amount of energy associated with such continuously random fluctuations.

We phrase the question in terms of an effective c-number field, $A_\mu^{\text{vac}}(x)$, whose net polarization vector can only lie in the "time" direction, $\mu = 4$, because of the random orientations of the fluctuating planes, as imagined by every observer in his own Lorentz frame. And we find that each such observer imagines the same form and magnitude of this vacuum energy. Note that the word "imagines" is used, rather than the more usual "sees" or "measures"; and this is because such an "effectively invariant'" vacuum energy cannot be described by conventional Lorentz transformations that are used to describe classical fields and particle transformations. This is an energy, an effective potential energy, that one cannot measure with ordinary equipment, especially since its frequency — the M value — is so very large. But if there is such an energy present, an average potential energy of all the fluctuating loops "stored" in the Quantum Vacuum, it can have significant effects on a large, classical scale, resisting the natural tendency of gravitational attraction.

And there is another difference between this $A_\mu^{\rm vac}(x)$ and conventional c-number fields, for it satisfies not an ordinary differential or integral equation, as might be expected of such a function, but (in momentum space) it corresponds to a 'distribution', and not to an ordinary 'function'. Perhaps this is why it has gone unnoticed for more than a century. Nevertheless, we propose this new form of effective potential energy as a real effect, due to the quantum fluctuations of the quantized fields; and we offer it as a possible example of the source of the present-day acceleration of the Universe. Because the shape of this effective potential energy at small distances (or times) is so suggestive of Inflation, whose treatment involves a significant variation from that provided in the present Chapter, but whose essence is of the same origin, the subject of Inflation will be treated in Chapter 13.

One prejudice of tho author should be stated at the outset, which forms part of the motivation for the present remarks. A conventional approach to vacuum energy is that the latter represents, in some fashion, zero point energies of relevant quantum fields. Aside from being horribly divergent, and so requiring massive renormalizations, it is not clear that zero point terms even belong in any field theory, for they can be well understood as the remnants of improper positioning of products of operators at the same space-time point. One cannot proceed from classical to quantum forms without facing this question, which has long ago been given a Lorentz invariant answer by [Symanzik (1961)] in terms of "normal ordering". When Lorentz variance is subsumed into a more general relativistic invariance, in which all energies couple to the metric, there is still no requirement of including those zero point terms which should have been excluded at the very beginning; instead, zero point terms have been pressed into service as the simplest way of attempting to understand vacuum energy.

Because zero point energies have been used to give a qualitative explanation of the logarithm of the lowest order Lamb shift does not mean that they are the correct explanation for that effect; zero point energies cannot produce the additive constant to that logarithm, and they most certainly cannot produce the higher order terms which have been experimentally verified. It is here suggested that there is another source of vacuum energy, one which violates no principles of quantum field theory, that can provide a simple and at least qualitatively-reasonable QED mechanism for both Inflation, occurring at the beginning of our Universe, and for its remnant Dark Energy, which continues that acceleration at a greatly reduced and gradually declining rate.

8.2 Formulation

We take the point of view that such an averaged $A_{\mu}^{\text{vac}}(x)$ can be treated as an effective, classical, external field, albeit one which is always present, and should be included in QED considerations, and specially in the construction of the QED Generating Functional (GF). There, the Schwinger-Symanzik-Fradkin formulations produce a GF, $Z[j, \eta, \bar{\eta}]$, which, for our purposes can be transformed into the most convenient form:

$$\mathcal{Z}[\eta, \bar{\eta}, j] = \mathcal{N} \exp\left[\frac{i}{2}\int d^4x d^4y j^{\mu}(x) D_{c,\mu\nu}(x-y) j^{\nu}(y)\right] \times e^{\mathcal{D}_A} \exp\left[i\int d^4x\, d^4y\, \bar{\eta}(x) G_c(x, y|A)\eta(y)\right] \exp[L[A]] \quad (8.1)$$

with $A_{\mu}(x) = \int d^4y D_{c,\mu\nu}(x-y) j^{\nu}(y)$.

The relativistic notation used throughout this article is the so–called "east coast metric", with the scalar product defined by $a \cdot b = \vec{a} \cdot \vec{b} - a_0 b_0$, and the Dirac matrices such that $\gamma_{\mu}^{\dagger} = \gamma_{\mu}$, $\gamma_{\mu}^2 = 1$ and $\{\gamma_{\mu}, \gamma_{\nu}\} = 2\delta_{\mu\nu}$.

The normalization $\mathcal{Z}[0,0,0] = 1$ is defined by: $\mathcal{N}^{-1} = \langle 0|S|0\rangle = e^{\mathcal{D}_A} \exp[L[A]]\Big|_{A=0}$, $G_c[A] = [m + \gamma \cdot (\partial - ieA)]^{-1}$ is the Feynman (casual) electron propagator in an arbitrary external field $A_{\mu}(x)$; $L[A] = Tr \ln(G_c^{-1}[A] G_c[0])$ is the vacuum functional corresponding to a single closed lepton loop, to which is attached the sum of all possible (even) number of fields A_{μ}; and $D_c(x-y)$ is the free photon propagator in an arbitrary relativistic gauge. We shall again refer to the quantity:

$$e^{\mathcal{D}_A} \text{ with } \mathcal{D}_A = -\frac{i}{2}\int d^4x\, d^4y \frac{\delta}{\delta A_{\mu}(x)} \mathcal{D}_{c,\mu\nu}(x-y) \frac{\delta}{\delta A_{\mu}(y)}$$

as the linkage operator, for its function is to link all pairs of A-dependence upon which it acts by virtual D_c propagators. We begin this discussion specifying those radiative corrections due to a virtual electron-positron bubble, but will later generalize to the corresponding situations when similar contributions are made by virtual bubbles of other particles.

Equation (8.1) as written is true when no classical, external field is present; however if such fields were present, (8.1) would still be true if the A_μ in G_c and L were replaced by $A_\mu + A_\mu^{\text{ext}}$. In this way, it is easy to see the relation between QED and Quantum Mechanics: were the A dependence and the linkage operator suppressed, one has the GF for many-body Potential Theory, where the fermions are moving in the potential A_μ^{ext}. All of the virtual structure of QED corresponds to the action of that linkage operator, connecting all the A dependence upon which it acts, in all possible ways.

Let us now assume the existence of an effective, classical field arising from the virtual fluctuations of the QED vacuum; for the present discussion, no ordinary, classical, external field need be included. Conventionally, the vacuum expectation value (vev) of the current operator $j_\mu(x) = ie\bar{\psi}(x)\gamma_\mu\psi(x)$ must vanish in the absence of classical, external fields, designated by $A_\mu^{\text{ext}}(x)$:

$$< 0|j_\mu(x)|0 > \Big|_{A^{\text{ext}}=0} = 0.$$

When such a classical A_μ^{ext} is present, the current it induces in the vacuum can be non-zero:

$$< 0|j_\mu(x)|0 > \Big|_{A^{\text{ext}}\neq 0} \neq 0$$

although strict current conservation demands that

$$\partial_\mu < 0|j_\mu(x)|0 > \Big|_{A^{\text{ext}}\neq 0} = 0.$$

The conventional mathematical apparatus used to describe the vacuum state, or vev of current operators, makes no reference to the scales on which vacuum properties are to be observed; and in this sense, it is here suggested that the conventional description is incomplete. On distance scales larger than the electron's Compton wave length, $\lambda_e \sim 10^{-10}$ cm, one can imagine that the average separation of virtual e^+ and e^- currents is not distinguishable, and hence that the vacuum displays not only a zero net charge, but also a zero charge density. But on much smaller scales, such as 10^{-20} cm, the average separation distances between virtual e^+ and e^- are relatively large, with such currents describable as moving in each other's fields until they annihilate. Since there is nothing virtual, or "off shell", about charge, on sufficiently small space-time scales, such "separated" currents can be imagined to produce "effective c-number" fields, characterized by an $A_\mu^{\text{vac}}(x)$, which could not be expected to be measured at distances larger than λ_e, but which exist and contain electromagnetic energy on scales much smaller than λ_e. We therefore postulate that, at such small scales and in the absence of conventional, large scale A_μ^{ext}, $< 0|j_\mu(x)|0 >$ need be neither x-independent nor zero; but rather, that it generates an $A_\mu^{\text{vac}}(x)$ discernible at such small scales, which is given by the conventional expression:

$$A_\mu^{\text{vac}}(x) = \int d^4\, y\, D_{c,\mu\nu}(x-y) < 0|j_\nu(y)|0 > \tag{8.2}$$

where $D_{c,\mu\nu}$ is the usual, free field, Feynman (causal) photon propagator that, for convenience, is defined in the Lorentz gauge, $0 = \partial_\mu A_\mu^{\text{vac}} = \partial_\mu D_{c,\mu\nu} = \partial_\nu D_{c,\mu\nu}$.

For comparison, note that classical electromagnetic vector potentials can always be written in terms of well-defined, classical currents J_μ, by an analogous relation: $A_\mu = \int D_{c,\mu\nu}(x-y)J_\nu$, while the transition to operator QED in the absence of conventional, large scale, external fields involves the replacements of the classical vector potential and currents by operators $A_\mu(x)$ and $j_\mu(x) = ie\bar{\psi}(x)\gamma_\mu\psi(x)$, which satisfy operator equations of motion: $(-\partial^2)A_\mu(x) = j_\mu(x)$, or:

$$A_\mu(x) = \int D_{c,\mu\nu}(x-y)j_\nu(y) + \bar{A}_\mu(x) \tag{8.3}$$

where $\hat{A}$ denotes a free field operator satisfying $(-\partial^2)\hat{A} = 0$; its vev is zero.

Hence, calculating the vev of (8.3) yields:

$$< 0|A_\mu(x)|0 >= \int d^4y D_{c,\mu\nu}(x-y) < 0|j_\nu(y)|0 > \tag{8.4}$$

and conventionally, in the absence of the usual, large scale external fields, both sides of (8.4) are to vanish. However, if we assume that non zero $< 0|j_\mu(x)|0 >$ can exist on ultra short scales, then a comparison of (8.4) with (8.2) suggests that the $A_\mu^{\text{vac}}(x)$ produced by such small scale currents are to be identified with $< 0|A_\mu(x)|0 >$ found in conventional QED in the presence of the same $A_\mu^{\text{vac}}(x)$. In other words,

$$\begin{aligned} A_\mu^{\text{vac}}(x) = < 0|A_\mu(x)|0 >= \frac{1}{i}\int d^4y D_{c,\mu\nu}(x-y)\frac{\delta}{\delta A_\nu(y)} e^{-\frac{i}{2}\int \frac{\delta}{\delta A} D_c \frac{\delta}{\delta A}} \\ \times < 0|S[A^{\text{vac}}]|0 >^{-1} \exp[L[A + A^{\text{vac}}]]\Big|_{A\to 0}, \end{aligned} \tag{8.5}$$

which provides a bootstrap equation with which to determine such short scale $A_\mu^{\text{vac}}(x)$, if any exist. In (8.5), which can be transformed into a functional integral relation, the vacuum to vacuum amplitude is given by [4]:

$$< 0|S[A^{\text{vac}}]|0 >= e^{-\frac{i}{2}\int \frac{\delta}{\delta A} D_c \frac{\delta}{\delta A}} \exp[L[A + A^{\text{vac}}]]\Big|_{A\to 0} . \tag{8.6}$$

Of course, one immediate solution to (8.5) is $A_\mu^{\text{vac}}(x) = 0 =< 0|A_\mu(x)|0 >$, the conventional solution. But we are interested in, and shall find, solutions that may he safely neglected at conventional nuclear and atomic distances, but which are non-zero in an interesting way at much smaller distances. In a sense such non-zero solutions are akin to those found in symmetry breaking processes, such as spontaneous or induced magnetization; qualitatively similar ideas have previously been discussed elsewhere[1] for other reasons.

8.3 Approximation

How does one go about finding a solution to (8.5)? The first requirement is a representation for $L[A+A^{\text{vac}}]$ which is sufficiently transparent to allow the functional operation of (8.5) to be performed. Use of the Fradkin functional representation for $L[A]$ is one such possibility, because that representation is Gaussian in A, and (8.5) can be performed immediately. But one is then left with the task of evaluating the remaining functional operations, which is not a trivial affair. What we shall therefore do is to replace $L[A]$ by it's well-known, gauge-independent, second-order perturbative approximation, as derived in Chapter 6,

$$L[A + A^{\text{vac}}] \to \frac{i}{2}\int d^4x d^4y [A_\mu(x) + A_\mu^{\text{vac}}(x)] K_{\mu\nu}(x-y)[A_\nu(y) + A_\nu^{\text{vac}}(y)] \tag{8.7}$$

which is clearly Gaussian in A, and corresponds to the simplest Feynman diagram of a virtual, lepton-antilepton bubble. The strictly gauge-invariant form of that bubble may be represented as

$$\tilde{K}_{\mu\nu}^{(2)}(k) = (\delta_{\mu\nu}k^2 - k_\mu k_\nu)\prod^{(2)}(k^2), \quad k^2 = \vec{k}^2 - k_0^2,$$

with the renormalized form of $\prod^{(2)}(k^2)$, given as in (6.15) by

[1] [Guralnik (1964)] and other references quoted therein. A somewhat different but related approach can be found in the Conference Proceedings by [Rafelski et al. (2009)].

$$\prod_R^{(2)}(k^2) = \frac{2\alpha}{\pi}\int_0^l dy \cdot y(1-y)\ln\left[1 + y(1-y)\frac{k^2}{m^2}\right]. \tag{8.8}$$

Higher-order perturbative terms should each yield a less-important contribution to the final answer, although the latter could be qualitatively changed by their sum; we shall assume that this is not the case, and that the second-order, perturbative approximation - properly unitarized by the functional calculation which automatically sums over all such simple bubbles - gives a qualitatively reasonable approximation.

With (8.7) and (8.8), the functional operation of (8.5) is immediate, and yields the approximate equation for this A_μ^{vac},

$$A_\mu^{\text{vac}}(x) = \int d^4y \left(DK\frac{1}{1-DK}\right)_{\mu\nu}(x-y)A_\mu^{\text{vac}}(y), \tag{8.9}$$

or

$$\tilde{A}_\mu^{\text{vac}}(k) = \Pi_R^{(2)}(k^2)\left(\frac{1}{1\prod_R^{(2)}(k^2)}\right)\tilde{A}_\mu^{\text{vac}}(k). \tag{8.10}$$

The simplest non-zero solution to (8.10) may be found in the "tachyonic" form:

$$\tilde{A}_\mu^{\text{vac}}(k) = C_\mu(k)\ \delta(k^2 - M^2) = C_\mu(k)\ \delta(\vec{k}^2 - k_0^2 - M^2), \tag{8.11}$$

which then requires that

$$1 = \Pi_R^{(2)}(M^2)\frac{1}{1-\Pi_k^{(2)}(M^2)}, \quad \Pi_R^{(2)}(M^2) = \frac{1}{2}, \tag{8.12}$$

and which serves to determine M. Note that a solution of the form $C_\mu\delta(k^2 + \mu^2)$ would not be possible, since the log of $\prod_R^{(2)}(-\mu^2)$ picks up an imaginary contribution for time-like $k^2 = -\mu^2$, for $\mu > 2m$; $\prod_R^{(2)}(M^2)$ must be a real quantity to satisfy (8.10).

8.4 Computation

In order to fully describe this A_μ^{vac}, one must define $C_\mu(k)$. Remembering that the simplifying choice of a Lorentz gauge has already been made for this vacuum potential, one might expect to be able to choose:

$$C_\mu(k) = v_\mu - k_\mu(k_\nu v_\nu)/k^2$$

where v_μ is a constant to be determined. But the part proportional to k_μ is a pure gauge term, which cannot contribute to any electromagnetic field, and is therefore irrelevant, while the replacement of $C_\mu(k)$ by v_μ serves to generate fields that diverge in the region of the light cone. This is the form of solution which requires an *ad hoc* cut off when computing energies.

A far better solution is obtained by enforcing the Lorentz gauge condition with the replacement of $C_\mu(k)$ by $kv_\mu\delta(k\cdot v)$ where k is a constant to be determined; this choice produces a vacuum field that is both simple and everywhere finite. But before such a solution can be taken seriously, there are certain questions that must be answered:

(a) What is v_μ? Physically that vector should represent an electric field polarization, corresponding to the fluctuating electric field in the plane of the fluctuating bubbles. But the QED vacuum will have bubbles fluctuating in all possible planes, or even on curved surfaces; and there can be no $\vec{v}$ direction singled out. It is then intuitively clear that v_μ should have only a fourth component.

(b) But in which Lorentz frame? If it is to represent a field generated by the same vacuum processes in every frame, it should have the same value to each observer in his own Lorentz frame.

We now show that this choice of solution for the vacuum field does satisfy both of these requirements, and has just the correct behavior to suggest a mechanism for inflation, while producing a present day energy density that can be associated with Dark Energy.

Insert a representation for both delta functions of:

$$\tilde{A}_\mu^{\rm vac}(k) = k\ v_\mu\delta(k\cdot v)\ \delta(k^2 - M^2) \tag{8.13}$$

and calculate the Fourier transform of $\tilde{A}_\mu^{\rm vac}$. It will be represented by the parametric, "proper time" integral:

$$A_\mu^{\rm vac}(x) = \frac{k}{(2\pi)^4}\sqrt{\frac{i\pi}{4}}\left(\frac{-iv_\mu}{\sqrt{v^2}}\right)\int_{-\infty}^{+\infty} ds\ s^{-3/2}\epsilon(s)e^{isM^2+iX^2/4s} \tag{8.14}$$

with $\epsilon(s) = \frac{s}{|s|}$, $x\cdot v = \vec{r}\cdot\vec{v} - x_0v_0$, and where $X^2 = x^2 - (x\cdot v)^2/v^2$; that integral is a Lorentz invariant quantity. The corresponding solution in another Lorentz frame, represented by a prime on x and a prime on v, will have exactly the same form.

Now, consider the solution (8.14) in our frame, and ask what value should be assigned to the spatial components. From the argument of (a) above, the only sensible value for this field A_μ^{vac} is the choice $\vec{v} = 0$. Then, the quantity X^2 reduces to r^2, and (4.2) can be evaluated trivially:

$$A_\mu^{\text{vac}}(x) \to A_4^{\text{vac}}(x) = i\frac{k}{(2\pi)^3}\epsilon(v_0)\frac{\sin(Mr)}{r} \tag{8.15}$$

and depends only on r and on the sign of v_0.

Now switch to another Lorentz frame, where the result is given by (8.14) using primed variables, related to the unprimed variables by standard Lorentz transformation. An observer in that frame asks what value he should assign to the spatial components of his v'; and for his vacuum field, he comes to exactly the same conclusion as did we: the only sensible choice for a vacuum field as seen by him must require $\vec{v}' = 0$. Note that his square root variable $\sqrt{X^2}$ containing all the x' dependence is a Lorentz invariant quantity, and is equal to our square root variable. In our frame, when we set $\vec{v} = 0$, that variable reduces to r; and in his frame, when he sets $\vec{v}' = 0$ it reduces to r'. But both must be equal, since they were derived from the same invariant; when the observer in the primed frame sets his $\vec{v}' = 0$, as he must to describe his vacuum field, he is using the same functional form as (4.3) in terms of his r'.

The only possible difference between the two expressions of (8.15), primed or unprimed, is the sign of v_0 and that of v_0'. But, as always when dealing with physical entities, we restrict all admissible Lorentz transformations to those which are orthochronous, keeping the same sense of time, or of energy, or in this case of v_0; and hence $\epsilon(v_0) = \epsilon(v_0')$ and the two versions of (8.14) are the same. In this way, observers in every Lorentz frame see the same vacuum field. For simplicity, we shall choose $\epsilon(v_0) = +1$, although this choice of sign has no bearing on the vacuum energy densities to be calculated.

Following the arguments above, any observer in any frame will imagine a "vacuum electrostatic" potential of form:

$$\phi^{\text{vac}}(r) = k'\frac{\sin(Mr)}{r}, \quad k' = \frac{k}{(2\pi)^3}$$

where k' is a constant to be determined.

The resulting electric field has a rapid spatial variation, as does its energy density:

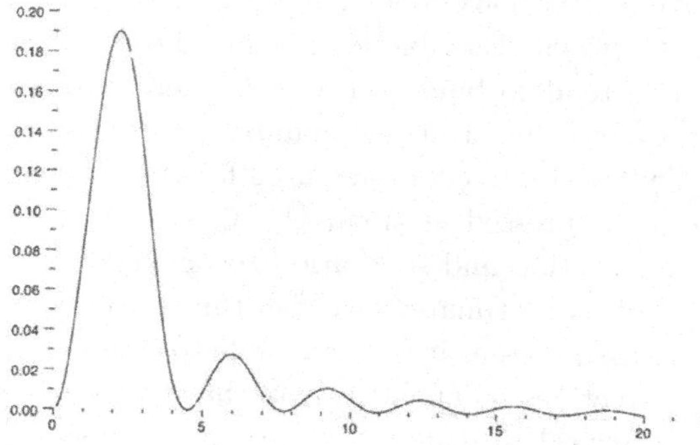

Fig. 8.1 A plot of $f(x) = \frac{1}{x^2}(\cos x - \frac{\sin x}{x})^2$ vs. x.

$$\rho = \vec{\mathcal{E}}^2/8\pi = \hat{\xi}\frac{M^2}{r^2}\left(\cos(Mr) - \frac{\sin(Mr)}{Mr}\right)^2 = \hat{\xi}M^4 f(x) \tag{8.16}$$

where $\hat{\xi} = k^{r2}/8\pi$, $x = Mr$, and $f(x) = \frac{1}{x^2}\left(\cos x - \frac{\sin x}{x}\right)^2$. A plot of $f(x)$ has the form indicated in Figure 8.1 which suggests the basic idea of this approach: the first pulse serves to kick-start Inflation, which is supposed to begin at $t \sim 10^{-42}$ s, and have an average energy density ρ such that $\rho^{1/4} \sim 10^{18}$ GeV. These numbers, and the time when inflation stops, $t \sim 10^{-32\pm6}$ s, with an average $\rho^{1/4} \sim 10^{13\pm3}$ GeV, are from Table 2.1 of [Liddle and Lyth (2000)]. However, the initial $\rho^{1/4}$ has simply been specified as the Planck mass, with no uncertainties attached, for the Planck mass just symbolizes the beginning of inflation; in reality, several orders of magnitude of uncertainties should be associated with that number, or with any such number for which a model exists.

8.5 Application to Dark Energy

The next question to be decided is which are the fundamental, quantized fields of Physics whose quanta are electrically charged, and are appropriate to be included. The three leptons and the quarks - of three colors and six flavors - are the natural, fundamental fermions out of which the charged particles forming the Standard Model are built; and it is these particles whose simplest closed-loops will form the A_μ^{vec} relevant to present-day Dark Energy.

One may have reservations about including such quark bubbles because of the strong interactions, describable in terms of flux tubes or gluon "bundles", which would tend to bind each $Q-\bar{Q}$ more strongly than that expected electromagnetically. But the counter argument to this is the observation that asymptotic freedom arranges for the attractive, multi-gluon interactions to be suppressed at small $Q-\bar{Q}$ separations, permitting the overall control of duration and separation to be electromagnetic. The author is not aware of any estimates in which this problem has been studied; and will therefore assume that it is permissible to consider such electromagnetic radiative corrections independently of the quarks' strong interactions, which become quite weak at distances less than an inverse pion wavelength.

In particular, we shall take into account fractional quark charges by multiplying each quark contribution by factors of either 4/9 or 1/9. We simplify the analysis by simplifying the y-dependence within the logarithms, and replace the average of the three lepton masses and of the six flavors of quark masses, by 1(GeV); these approximations are for convenience and simplicity only, and have no real bearing on the results. The relevant equation determining M is then

$$\frac{1}{2}=I_R(M^2)=\left(\frac{2\alpha}{3\pi}\right)\left\{3\ln\left(\frac{M}{\bar{m}_\ell}\right)+3\cdot 3\cdot\left(\frac{4}{9}+\frac{1}{9}\right)\ln\left(\frac{M}{\bar{m}_q}\right)\right\} \tag{8.17}$$

which, with $\bar{m}_\ell=\bar{m}_q\simeq 1$ GeV and $\alpha=1/137$, leads to $M\simeq 10^{18}$ GeV. It is interesting to note that this perfectly finite calculation, in the context of QED, is able to produce a "mass term" on the order of the Planck mass.

With this value of M, we are then able to compute the order-of-magnitude of present-day, vacuum energy density. If we compute that energy density by integrating the energy density of this vacuum field, as expressed in (8.16) — or more simply, by extracting the dominant, non-oscillatory behavior of that integral — and dividing by the present-day volume of the Universe, $R^3_{\text{now}}(4\pi/3)$, one finds

$$\rho_{\text{now}}\sim 6\hat{\xi}\cdot\frac{M^2}{R^2_{\text{now}}}\to 6\hat{\xi}\left(\frac{M}{m}\right)^2\left(\frac{mc}{\hbar}\right)\frac{m}{R^2_{\text{now}}} \tag{8.18}$$

where $m\simeq 10^{(-27)}$ gm is the electron mass, $(\hbar/mc)\simeq 10^{(-10)}$ cm, $\frac{M}{m}\simeq 5\times 10^{21}$, and $(4\pi'/3)R^3_{\text{now}}\simeq 10^{(85)}cm^3$, or $R^2_{\text{now}}\sim 10^{57}cm^2$. The result is

$$\rho_{\text{now}}\sim\hat{\xi}\cdot 10^{-30}gm/cm^3 \tag{8.19}$$

which, choosing $\hat{\xi}\sim 0(10)$ is the order of magnitude needed to fit the acceleration data [Kirshner (1999)][Reiss, et al. (2000)].

8.6 Partial Summary

This Model of a QED vacuum energy, defined by a particular choice of $C_\mu(k)$, is, of course, not the only possibility; but it is everywhere finite, with a form of Lorentz "invariance" more appealing than, e.g., a previous suggestion [Fried (2003)] for Dark Energy. And the large, first peak in Fig. 8.1, immediately suggests a possible connection with, and mechanism for Inflation, which will be discussed, in detail, in Part 4 of this Book. The basic QED vacuum bubble structure is maintained, but the quantized fields and their particles are somewhat different.

One may ask if such a Vacuum Energy can have experimentally observed effects on the many charged particles that compose our world. Two specific reasons answer this question in the negative; the first is that the frequency of such vacuum field oscillations is on the order of the Planck mass, and is much too rapid to be measured experimentally.

The second reason is somewhat deeper. This Vacuum Energy is due to the existence of fluctuations in the Quantum Vacuum. Although its energy density may still act to oppose gravitational attraction, this energy cannot be considered to be 'Potential' in the ordinary sense, possibly to be available for 'use' at any given time within the lifetime of our Universe, for then it would be diminished, or no longer exist. Rather, this Potential Energy must be understood to be contained, or restrained to lie within the Quantum Vacuum; and its possible appearance in the 'real world' could occur once, and only once, as suggested in Chapter 13, Section 3.

PART 3
Quantum Chromodynamics

Quantum Chromodynamics

We begin these Chapters by first noting that, by a simple re-arrangement, the Schwinger/Symanzik Generating Functional of QCD can be brought into a gauge-invariant form, even though the GF apparently contains gauge-dependent gluon propagators. This simple procedure has been overlooked for decades; and when combined with the Fradkin representations for the functionals $G[A]$ and $L[A]$, which are an integral part of the GF, as well as the most useful Halpern representation for $\exp[(-i/4)\int d^4xF^2]$, all relevant Gaussian functional operations — corresponding to the summation of all Feynman graphs of gluons exchanged between any pair of quarks and/or antiquarks — may be performed exactly. One then sees, explicitly, the cancelation of all gauge-dependent, gluon propagators; manifest gauge invariance (MGI) is here achieved as gauge-independence, and it appears in conjunction with a formalism containing manifest Lorentz covariance (MLC).

This cannot be done for an Abelian theory, but is possible for a non-Abelian theory which contains cubic and quartic gluon interactions, the very interactions which make a conventional perturbation expansion past the the first few orders a horrendous task. One also sees the appearance of a new and exact property called Effective Locality (EL), which simplifies all calculations by, in effect, transforming the remaining functional integrals into sets of ordinary integrals.

Further, it readily becomes apparent that, in this non-perturbative formulation, one cannot continue to consider quarks in the same, conventional fashion as quanta of other (Abelian) fields, such as electrons which satisfy the standard measurement properties of quantum mechanics: perfect position dependence at the cost of unknown momenta, and vice-versa. This is impossible for quarks since they always appear asymptotically in bound states, and their transverse coordinates can never, in principle, be measured exactly. Violation of this principle produces, in these non-perturbative solutions, absurdities in the exact evaluation of each and every QCD amplitude. To avoid this difficulty, a simple measure of transverse imprecision is phenomenologically introduced into the basic QCD Lagrangian, with two parameters that are essential to and determined from quark binding into hadrons. All previous absurdities of all now "realistic" QCD amplitudes are then removed, and one finds a simple method for constructing a quark-binding potential, which subsequently suggests an obvious generalization for the construction of nucleon-binding potentials.

This material is divided into a sequence of Chapters each containing and emphasizing one or more particular aspects of the formulation. In Chapter 9, the rearrangement alluded to above is discussed in detail, and the demonstration of rigorous gauge-invariance is presented, along with a brief description of Effective Locality. The need for Transverse Imprecision is demonstrated in Chapter 10, along with a simple phenomenological realization of that property, while Chapter 11 describes an obvious and very simple method of obtaining Quark-Binding Potentials, and simple estimates for pion and nucleon masses.

It should be emphasized that, for clarity and simplicity of presentation, all of this analysis has been performed for one type of massive SU(3) quark, and its complement of eight massless gluons; flavors, and weak and electromagnetic interactions, as well as quark spin, can easily be added, as desired, and one will then expect small changes in the parameters so estimated. The arguments presented here are "proof of principle" arguments, which show just how non-perturbative, gauge-invariant solutions for QCD amplitudes may be obtained.

Chapter 12 extends the quark-binding treatment of the previous Chapter for the construction of nucleon-binding potentials, of which the simplest example is a "model deuteron" whose binding potential turns out to be qualitatively similar to those assumed years ago, on the basis of physical arguments. Again, with the neglect of quark and nucleon spin, and electromagnetic interactions, which can all be added as desired, we are able to bind two "model nucleons" with a derived potential that is physically most reasonable. And that potential, along with the obvious generalizations that it immediately suggests, may well form the basis for understanding the nucleon-building of heavier nuclei, resulting from basic, realistic QCD.

Finally, mention should be made of the relative simplicity of this approach, compared to other well-known and long-studied methods of calculation in QCD. Again, it is the unexpected EL, appearing automatically after the non-perturbative sum over all possible, gauge-invariant, gluon exchanges between quarks has occurred, which is responsible for the huge simplifications obtained as Halpern's functional integrals are here reduced to a finite set of ordinary integrals amenable to computer evaluation, and here estimated in the simplest way possible. One has long believed in the Principle of *"Conservation of Difficulty"*, when calculating higher-order effects in QED or any Abelian theory; but for non-Abelian field theories, approached in the manner suggested here, that Principle is not true.

Chapter 9

Explicit, Non-Perturbative Gauge Invariance

9.1 From QED to QCD

It is simplest to begin with QED, and its free-photon Lagrangian,

$$\mathcal{L}_0 = \frac{1}{4} f_{\mu\nu} f\mu\nu = -\frac{1}{4}(\partial_\mu A_\nu - \partial_\nu A_\mu)^2, \tag{9.1}$$

whose action integral can be rewritten as

$$\begin{aligned}\int d^4 \mathcal{L}_0 &= -\frac{1}{2}\int (\partial_\nu A_\mu)^2 + \frac{1}{2}\int (\partial_\mu A_\mu)^2 \\ &= -\frac{1}{2}\int A_\mu(-\partial^2)A_\mu + \frac{1}{2}\int (\partial_\mu A_\mu)^2.\end{aligned} \tag{9.2}$$

The difficulty of maintaining both manifest gauge invariance (MGI) and manifest Lorentz covariance (MLC) appears at this stage. What has typically been done is to use the last right-hand-side (R.H.S.) term of (9.2) to define a relativistic gauge in which all calculations maintain MLC, while relying on strict charge conservation to maintain an effective gauge invariance of the theory. The choice of relativistic gauge can be arranged in various ways; the simplest functional way is to multiply the last RHS term of (9.2) by the real parameter λ, and treat it as an "interaction" term. With the free-field, ($\lambda = 0$, Feynman) propagator $D^{(0)}_{c,\mu\nu} = \delta_{\mu\nu} D^{(0)}_c$, where $(-\partial^2)D_C = 1$, the free-field photon GF is given by

$$\mathcal{Z}^{(0)}[j] = \exp\left[\frac{i}{2}\int j \cdot D^{(0)}_c \cdot j\right]. \tag{9.3}$$

Then, operating upon it by the "interaction" λ-term, a new, free-field GF is produced,

$$\begin{aligned}\mathcal{Z}_0^{(\zeta)}[j] &= e^{\frac{1}{2}\lambda\int(\partial_\mu A_\mu)^2}\Big|_{A\to\frac{1}{i}\frac{\delta}{\delta j}}\cdot e^{\frac{i}{2}\int j\cdot D_c^{(0)}j} \\ &= e^{\frac{i}{2}\int j\cdot D_c^{(\zeta)}\cdot j}\cdot e^{\frac{1}{2}\mathbf{Tr}\ln[1-\lambda\partial\otimes\partial/\partial^2]},\end{aligned} \tag{9.4}$$

where $D_{c,\mu\nu}^{(\zeta)} = [\delta_{\mu\nu} - \zeta\partial_\mu\partial_\nu/\partial^2]D_c$, with $\zeta = \lambda/(1-\lambda)$. The functional "linkage operation" of (9.4) is fully equivalent to a bosonic, Gaussian functional integration (FI), and is frequently more convenient than a standard FI; in other, related situations, it can be more illuminating than the conventional FI.

The Tr-Log term of (9.4) is an infinite phase factor, representing the sum of vacuum energies generated by longitudinal and time-like photons, with a weight λ arbitrarily inserted; this quantity can be removed by an appropriate version of normal ordering, but can more simply be absorbed into an overall normalization constant.

Including the conventional fermion interaction, $L_{\text{INT}} = -ig\bar{\varphi}\gamma\cdot A\psi$, and the above "gauge interaction" term, $\frac{1}{2}\lambda(\partial_\mu A_\mu)^2$, one then writes the standard Schwinger/Symanzik solution for the GF in this gauge as

$$\mathcal{Z}_{\text{QED}}^{(\zeta)}[j,\eta,\bar{\eta}] = Ne^{i\int\bar{\eta}G_c[A]\eta + L[A] + \frac{i}{2}\lambda\int(\partial_\mu A_\mu)^2}\Big|_{A\to\frac{1}{i}\frac{\delta}{\delta j}}\cdot e^{\frac{i}{2}\int j\cdot D_c^{(0)}\cdot j}, \tag{9.5}$$

using the methods of Chapter 3, where

$$G_c[A] = [m+\gamma\cdot(\partial - igA)]^{-1}, \quad L[A] = \mathbf{Tr}\ln[1 - ig\gamma\cdot AS_c], \quad S_c = G_C[0]$$

and where the phase factor of (9.4) has been absorbed into N. A most convenient re-arrangement (9.5) uses the easily-proven identity of Chapter 2 for an arbitrary function $\mathcal{F}[A]$,

$$\mathcal{F}\left[\frac{1}{i}\frac{\delta}{\delta j}\right]e^{\frac{i}{2}\int jD_c^{(\zeta)}j} \equiv e^{\frac{i}{2}\int j\mathcal{D}_c^{(\zeta)}j}\cdot e^{\mathcal{D}_A^{(\zeta)}}\cdot e^{i\int\bar{\eta}G_c[A]\eta + L[A]}\Big|_{A=\int D_c^{(\zeta)}j}, \tag{9.6}$$

where $\exp[\mathcal{D}_A^{(\zeta)}]$ denotes the relevant "linkage operator", with

$$\mathcal{D}_A^{(\zeta)} = -\frac{i}{2}\int\frac{\delta}{\delta A}\cdot D_c^{(\zeta)}\cdot\frac{\delta}{\delta A},$$

so that (9.5) now reads

$$\mathcal{Z}_{\text{QED}}^{(\zeta)}[j,\eta,\bar{\eta}] = N e^{\frac{i}{2}\int jD_c^{(\zeta)}j} \cdot e^{\mathcal{D}_A^{(\zeta)}} \cdot e^{i\int \bar{\eta}G_c[A]\eta + L[A]}\bigg|_{A=\int D_c^{(\zeta)}j}. \tag{9.7}$$

This is the formal solution for the GF of QED in the gauge ζ which has been known and used for more than a half-century.

We now come to QCD, with

$$\mathcal{L}_{\text{QCD}} = \frac{1}{4}\mathbf{F}_{\mu\nu}^a \mathbf{F}_{\mu\nu}^a - \bar{\psi}\cdot[m + \gamma_\mu(\partial_\mu - igA_\mu^a\lambda^a)]\cdot\psi, \tag{9.8}$$

where $\mathbf{F}_{\mu\nu}^a = \partial_\mu A_\nu^a - \partial_\nu A_\mu^a + gf^{abc}A_\mu^b A_\nu^c \equiv \mathbf{f}_{\mu\nu}^a + gf^{abc}A_\mu^b A_\nu^c$, and we now set up QCD in the form used above for QED. As a final preliminary step, we write

$$-\frac{1}{4}\int F^2 = -\frac{1}{4}\int f^2 - \frac{1}{4}\int [F^2 - f^2] = -\frac{1}{4}\int f^2 + \int \mathcal{L}'[A], \tag{9.9}$$

with $\mathbf{f}_{\mu\nu}^a = \partial_\mu A_\nu^a - \partial_\nu A_\mu^a$ and $\mathcal{L}'[A] = -\frac{1}{4}(2\mathbf{f}_{\mu\nu}^a + gf^{abc}A_\mu^b A_\nu^c)(gf^{abc}A_\mu^b A_\nu^c)$, and for subsequent usage, after an integration-by-parts, we note the exact relation

$$-\frac{1}{4}\int \mathbf{F}^2 = -\frac{1}{2}\int A_\mu^a(-\partial^2)A_\mu^a + \frac{1}{2}\int (\partial_\mu A_\mu^a)^2 + \int \mathcal{L}'[A]. \tag{9.10}$$

(In the next few paragraphs, for simplicity, we suppress the fermion/quark variables, which will be re-inserted at the end of this discussion.)

In order to select a particular relativistic gauge, one can multiply the second R.H.S. term of (9.10) by λ, and include this term as part of the interaction, thereby obtaining the familiar QCD GF in the relativistic gauge specified by

$$\mathcal{Z}_{\text{QCD}}^{(\zeta)}[j] = \mathcal{N} e^{i\int \mathcal{L}'[\frac{1}{i}\frac{\delta}{\delta j}]} \cdot e^{\frac{i}{2}\lambda\int \frac{\delta}{\delta j_\mu}\partial_\mu\partial_\nu\frac{\delta}{\delta j_\nu}} \cdot e^{\frac{i}{2}\int j\cdot D_c^{(0)}\cdot j}, \tag{9.11}$$

or, after re-arrangement,

$$\mathcal{Z}_{QCD}^{(\zeta)}[j] = \mathcal{N} e^{i\int \mathcal{L}'[\frac{1}{i}\frac{\delta}{\delta j}]} \cdot e^{\frac{i}{2}\int j\cdot \mathbf{D}_c^{(\zeta)}\cdot j}, \tag{9.12}$$

with the determinantal phase factor of (9.4) included in the normalization $\mathcal{N}$, and a δ^{ab} associated with each free-gluon propagator $\mathbf{D}_c$.

After re-inserting the quark variables, and after re-arrangement, expansion of (9.12) in powers of g clearly generates the conventional Feynman graphs of perturbation theory in the gauge ζ. It is clear that all choices of λ are possible except $\lambda = 1$, for that choice leads to $\zeta \to \infty$ and an ill-defined gluon propagator. This is an unfortunate situation, because the choice $\lambda = 1$ is precisely the one that corresponds to MGI in QCD, as is clear from (9.10).

But there is a very simple way of re-writing (9.12), by replacing the $\int \mathcal{L}'[A]$ of that equation by the relation given by (9.10),

$$i \int \mathcal{L}'[A] = -\frac{i}{4} \int \mathbf{F}^2 + \frac{i}{2} \int A^a_\mu (-\partial^2) A^a_\mu - \frac{i}{2} \int (\partial_\mu A^a_\mu)^2, \tag{9.13}$$

which (continuing to suppress the quark variables momentarily) yields

$$\mathcal{Z}^{(\zeta)}_{\mathrm{QCD}}[j] = \mathcal{N} e^{-\frac{i}{4}\int F^2 - \frac{i}{2}(1-\lambda)\int (\partial_\mu A^a_\mu)^2 + \frac{i}{2}\int A^a_\mu(-\partial^2)A^a_\mu}\Big|_{A \to \frac{1}{i}\frac{\delta}{\delta j}} \cdot e^{\frac{i}{2}\int j\cdot \mathbf{D}^{(0)}_c \cdot j}. \tag{9.14}$$

It is now obvious that the choice $\lambda = 1$ can be made. It will become clear below that, using (9.14), the functional operations effectively treat gluons as if they were quanta of a "ghost" field. The gluons of the theory, never measurable by themselves, disappear effectively from the exact calculation of every fermionic QCD correlation function, without approximation and without exception. This "ghost mechanism" occurs because all factors of

$$e^{\frac{i}{2}\int j\cdot \mathbf{D}^{(0)}_c \cdot j},$$

of (9.14), in the sum of all virtual gluon processes, are effectively removed by the action of the term

$$e^{\frac{i}{2}\int A^a_\mu(-\partial^2)A^a_\mu}\Big|_{A \to \frac{1}{i}\frac{\delta}{\delta j}},$$

of (9.14). In the end, the gluon acts as a "spark plug" to generate the MGI and MLC interactions of the theory, which then take on remarkably simple forms.

If one is interested in the radiative corrections to the gauge-dependent, free-field gluon propagator, whose corrections are now guaranteed to be gauge-independent, the leading R.H.S. factor of

$$e^{\frac{i}{2}\int j\cdot \mathbf{D}^{(0)}_c \cdot j}$$

should be retained in the rearranged expression of (9.14), taken at $\lambda = 1$,

$$\mathcal{Z}^{(0)}_{\mathrm{QCD}}[j] = \mathcal{N} e^{\frac{i}{2}\int j\cdot\mathbf{D}^{(0)}_c\cdot j} . e^{-\frac{i}{2}\int \frac{\delta}{\delta A}\cdot\mathbf{D}^{(0)}_c\cdot\frac{\delta}{\delta A}} . e^{-\frac{i}{4}\int F^2+\frac{i}{2}\int A\cdot(-\partial^2)\cdot A}\Big|_{A=\int \mathbf{D}_c(0)\cdot j} . \tag{9.15}$$

Otherwise, and for the specific examples of gluon exchanges between quark lines to follow, this factor plays no role and will be suppressed. The resulting GF is then MGI, and its superscript will be suppressed.

After re-inserting quark variables, (9.15) becomes

$$\begin{aligned}&\mathcal{Z}_{\mathrm{QCD}}[j,\bar{\eta},\eta]\\&= \mathcal{N} e^{-\frac{i}{2}\int \frac{\delta}{\delta A}\cdot\mathcal{D}^{(0)}_A\cdot\frac{\delta}{\delta A}} \cdot e^{-\frac{i}{4}\int \mathbf{F}^2+\frac{i}{2}\int A\cdot(-\partial^2)\cdot A} \cdot e^{i\int \bar{\eta}\cdot\mathbf{G}_c[A]\cdot\eta+L[A]}\Big|_{A=\int \mathbf{D}^{(0)}_c\cdot j}\end{aligned} \tag{9.16}$$

and we next invoke the representation suggested by Halpern [Halpern (1977)]

$$e^{-\frac{i}{4}\int F^2} = \mathcal{N}' \int d[\chi] e^{\frac{i}{4}\int(\chi^a_{\mu\nu})^2+\frac{i}{2}\int\chi^a_{\mu\nu}\mathbf{F}^a_{\mu\nu}}, \tag{9.17}$$

where

$$\int d[\chi] = \prod_i \prod_a \prod_{\mu\nu} \int d\chi^a_{\mu\nu}(w_i), \tag{9.18}$$

so that (9.17) represents a functional integral over the anti-symmetric tensor field $\chi^a_{\mu\nu}$. Following the standard definition, all spacetime is broken up into small cells of size δ^4 about each point w_i, and $\mathcal{N}'$ is a normalization constant so chosen such that the R.H.S. of (9.17) becomes equal to unity as $\mathbf{F}^a_{\mu\nu} \to 0$. In this way, the GF may be re-written as ($\mathcal{N}' \cdot \mathcal{N} = \mathcal{N}'' \to \mathcal{N}$),

$$\begin{aligned}&\mathcal{Z}_{\mathrm{QCD}}[j,\bar{\eta},\eta]\\&= \mathcal{N} \int d[\chi] e^{\frac{i}{4}\int\chi^2}\, e^{\mathcal{D}^{(0)}_A} \cdot e^{+\frac{i}{2}\int\chi\cdot\mathbf{F}+\frac{i}{2}\int A\cdot(-\partial^2)\cdot A}\\&\cdot e^{i\int \bar{\eta}\cdot\mathbf{G}_c[A]\cdot\eta+\mathbf{L}[A]}\Big|_{A=\int \mathbf{D}^{(0)}_c\cdot j},\end{aligned} \tag{9.19}$$

where $\exp[\mathcal{D}^{(0)}_A]$ is the linkage operator with $\mathcal{D}^{(0)}_A = -\frac{i}{2}\int \frac{\delta}{\delta A}\cdot\mathbf{D}^{(0)}_c\cdot\frac{\delta}{\delta A}$.

Quarks and anti-quarks are treated as stable entities during any scattering, production or binding process, which means that relevant functional derivatives with respect to the sources $\eta, \bar{\eta}$, will bring down factors of $\mathbf{G}_c(x, yIA)$, one such factor for each quark or antiquark under discussion. For example, by standard mass-shell amputation, one can then pass to the construction of a scattering amplitude of a pair of quarks, or of any number of quarks, or of a quark-antiquark pair. If this scattering is to occur at high-energies and small momentum transfer, a convenient and relatively simple eikonal approximation is available, derived in detail in Appendix B of [Avan, et al. (2003)]. This will be a most convenient tool in Chapter 11 where we employ the well-known connection between an eikonal function dependent upon impact parameter, and an effective potential which is the cause of the scattering or production or binding which leads to that eikonal function.

But it is worth emphasizing that any such, simplifying eikonal approximation is not to be confused with the exact functional representations corresponding to scattering, production and binding. Because the linkage operator in (9.19) represents an effective Gaussian functional operation upon the A-dependence contained within $\mathbf{G}_c[A]$ and $\mathbf{L}[A]$, and because there exist Fradkin representation of these functionals which are Gaussian in A, the specific functional operations required, resulting from well-defined functional derivatives with respect to the sources, $\eta, \bar{\eta}$, according to the physical process under consideration, can be performed exactly. This produces the sum of all Feynman graphs corresponding to the exchange of an infinite number of gluons between quarks and/or anti-quarks, exhibited in terms of the Fradkin functional parameters that define the representations for $\mathbf{G}_c[A]$ and $\mathbf{L}[A]$. Exact expressions of Fradkin representations for these non-Abalian functions are reproduced in Appendix D.

9.2 Gluon Summations and Explicit Gauge Invariance

The correlation functions of QCD are obtained by appropriate functional differentiation of (9.19) with respect to gluon and quark sources. Since we are here concerned only with quark (Q) or anti-quark ($\bar{Q}$) interactions, in which all possible numbers of virtual gluons are exchanged, we immediately set the gluon sources j equal to zero. All $Q/\bar{Q}$ amplitudes are then obtained by pairwise functional differentiation of the quark sources $\eta(y)$, $\bar{\eta}(x)$; and each such operation "brings down" one of a set of (properly anti-symmetrized) Green's functions, $\mathbf{G}_c[A]$. For example, the 2-point quark propagator will involve the FI $\int d[\chi]$ and the linkage operator acting upon $\mathbf{G}_c(x,y|A)\exp\{\mathbf{L}[A]\}$, followed by setting $A \to 0$. Similarly, the $Q/\bar{Q}$ scattering amplitude will be obtained from the same functional operations acting upon the (anti-symmetrized) combination $\mathbf{G}_c[A]\mathbf{G}_c[A]\exp\{\mathbf{L}[A]\}$, followed by $A \to 0$, as

$$\begin{aligned} &M(x_1,y_1;x_2,y_2) \\ &= \frac{\delta}{\delta\bar{\eta}(x_1)}\cdot\frac{\delta}{\delta\eta(y_1)}\cdot\frac{\delta}{\delta\bar{\eta}(x_2)}\cdot\frac{\delta}{\delta\eta(y_2)}\cdot Z_c\{0,\bar{\eta},\eta\}\Big|_{\eta=\bar{\eta}=0;j=0} \\ &= \mathcal{N}\int d[\chi]e^{\frac{i}{4}\int\chi^2}e^{\mathcal{D}_A^{(0)}}e^{+\frac{i}{2}\int\chi\cdot\mathbf{F}+\frac{i}{2}\int A\cdot(\mathbf{D}_c^{(0)})^{-1}} \\ &\quad \mathbf{G}_c(x_1,y_1|gA)\mathbf{G}_c(x_2,y_2|gA)e^{\mathbf{L}[A]}\Big|_{A=0} \\ &\quad - \{1 \leftrightarrow 2\}; \end{aligned} \tag{9.20}$$

and other fermionic $2n$-polnt functions are obtained in the same way.

The Fradkin functional representations for $\mathbf{G}_c[A]$ and $\mathbf{L}[A]$, derived in Appendix D display a Gaussian dependence on A, and hence the linkage operations of (9.20), in any order of the expansion of $\exp\{\mathbf{L}[A]\}$ in powers of $\mathbf{L}$, or somewhat more conveniently, using a functional cluster expansion for $\exp\{\mathcal{D}_A^{(0)}\}$ operating upon $\exp\{\mathbf{L}[A]\}$, can be performed exactly. One small complication, easily surmounted, is due to the non-Abelian nature of QCD, in which the A-dependence appears inside an ordered exponential, ordered in terms of an invariant "Schwinger proper time" variable, s, as in the explicit representations for $\mathbf{G}_c(x,y|A)$ and $\mathbf{L}[A]$. However cumbersome these representations may appear, the essential fact is that they are Gaussian in A, and hence the linkage operations, corresponding to the summation over all possible gluon exchanges between Q and/or $\bar{Q}$ lines, may be performed exactly. One then finds a simple, effectively local representations for the sum over all such exchanges, here called a "gluon bundle", or GB. Henceforth, "bundle-diagrams" will replace Feynman diagrams containing individually-specified gluon exchanges.

To display the gauge invariance of all such QCD correlation functions, one first notes that $\mathbf{L}[A]$ is explicitly invariant under arbitrary changes of the full QCD gauge group, as demonstrated in [Fried (1990)]. Then, it is convenient to combine the Gaussian A-dependence of very entering $\mathbf{G}_c[A]$ into the quantity

$$\exp\left[\frac{i}{2}\int d^4z A_a^\mu(z)\tilde{\mathbf{K}}_{\mu\nu}^{ab}(z)A_b^\nu(z) + i\int d^4z\,\tilde{\mathbf{Q}}_a^\mu(z)A_\mu^z(z)\right], \tag{9.21}$$

where $\tilde{\mathbf{K}}$ and $\tilde{\mathbf{Q}}$ are local functions of the Fradkin variables, collectively denoted by $u_\mu(s')$, and the $\Omega^a(s_1)$, $\Omega^a(s_2)$, $\Phi^a_{\mu\nu}(s_1)$ and $\Phi^a_{\mu\nu}(s_2)$ are needed to extract the $A^a_\mu(y-u(s'))$ from under ordered exponentials. Note that the $\tilde{\mathbf{Q}}$ and $\tilde{\mathbf{K}}$ are also to represent the sum of similar contributions from each of the $\mathbf{G}_c(x,y|A)$ which collectively generate the amplitude under consideration. For example, in the case of the 4-point function which one obtains from the product of $\mathbf{G}_{c,I}[A]$ and $\mathbf{G}_{c,II}[A]$,

$$\begin{aligned}\tilde{\mathbf{K}}^{ab}_{\mu\nu}(z) = 2g^2 \int_0^{s_1} ds'\delta^{(4)}(z-y_1+u(s'))f^{abc}\Phi^c_{\mu\nu,I}(s') \\ + 2g^2 \int_0^{s_2} ds'\delta^{(4)}(z-y_2+\bar{u}(s')f^{abc}\Phi^c_{\mu\nu,II}(s') \end{aligned} \tag{9.22}$$

and

$$\begin{aligned}\tilde{\mathbf{Q}}^a_\mu(x) = &-2g\partial_\nu\Phi^a_{\nu\mu,I}(z) - g\int_0^{s_1} ds'\delta^{(4)}(z-y_1+u(s'))u'_\mu(s')\Omega^a_I(s') \\ &-2g\partial_\nu\Phi^a_{\nu\mu,II}(z) - g\int_0^{s_2} ds'\delta^{(4)}(z-y_2+\bar{u}(s'))\bar{u}'_\mu(s')\Omega^a_{II}(s'),\end{aligned} \tag{9.23}$$

where subscripts 1, 2 and I, II are used interchangeably to denote particles 1 and 2; and, for clarity, $\bar{u}(s')$ is used to denote particle 2, and the function

$$\Phi^a_{\mu\nu}(z) \equiv \int_0^{s} ds'\delta^{(4)}(z-y+u(s'))\Phi^a_{\mu\nu}(s') \tag{9.24}$$

is introduced in $\tilde{\mathbf{Q}}$ for ease of presentation. For higher quark n-point functions, there will be additional terms contributing to $\tilde{\mathbf{Q}}$ and $\tilde{\mathbf{K}}$, but their forms will be the same.

Combining the quadratic and linear A-dependence from $\tilde{\mathbf{K}}$ and $\tilde{\mathbf{Q}}$ above, and including that $\mathbf{L}[A]$ dependence explicitly written in (9.19), the needed linkage operation reads

$$\exp\left[\frac{i}{2}\int\frac{\delta}{\delta A}\cdot\mathbf{D}^{(0)}_{\rm c}\cdot\frac{\delta}{\delta A}\right]\cdot\exp\left[\frac{i}{2}\int A\cdot\bar{\mathbf{K}}\cdot A + i\int\tilde{\mathbf{Q}}\cdot A\right]\cdot\exp(\mathbf{L}[A]), \tag{9.25}$$

where

$$\langle z|\bar{\mathbf{K}}^{ab}_{\mu\nu}|z'\rangle = [\tilde{\mathbf{K}}^{ab}_{\mu\nu}(z)+gf^{abc}\chi^c_{\mu\nu}(z)]\delta^{(4)}(z-z')+\langle z|(\mathbf{D}^{(0)}_{\rm c})^{-1}|^{ab}_{\mu\nu}|z'\rangle \tag{9.26}$$

and

$$\bar{\mathbf{Q}}^a_\mu(z) = \tilde{\mathbf{Q}}^a_\mu(z) + \partial_\nu \chi^a_{\mu\nu}(z). \tag{9.27}$$

In $\bar{\mathbf{K}}$, all terms but the inverse of the gluon propagator are local.

Equation (9.25) requires the linkage operator to act upon the product of two functionals of A, and this can be represented by the identity

$$e^{\mathcal{D}_A}\mathcal{F}_1[A]\mathcal{F}_2[A] = (e^{\mathcal{D}_A}\mathcal{F}_1[A])e^{\overleftrightarrow{\mathcal{D}}}(e^{\mathcal{D}_A}\mathcal{F}_2[A]), \tag{9.28}$$

where, with an obvious notation, the "cross-linkage" operator $\exp\{\overleftrightarrow{\mathcal{D}}\}$ is defined by

$$\overleftrightarrow{\mathcal{D}} = -i\int \frac{\overleftarrow{\delta}}{\delta A}\mathbf{D}_c^{(0)}\frac{\overrightarrow{\delta}}{\delta A}. \tag{9.29}$$

With the identifications

$$\mathcal{F}_1[A] = \exp\left[\frac{i}{2}\int A\cdot\bar{\mathbf{K}}\cdot A + i\int\bar{\mathbf{Q}}\cdot A\right], \quad \mathcal{F}_2[A] = \exp(\mathbf{L}[A]), \tag{9.30}$$

the evaluation of $e^{\mathcal{D}_A}\mathcal{F}_1[A]$ is given by a standard, functional identity of Chapter 2, and reads

$$\begin{aligned} e^{\mathcal{D}_A}\mathcal{F}_1[A] = &\exp\left[\frac{i}{2}\int\bar{\mathbf{Q}}\cdot\mathbf{D}_c^{(0)}\cdot(1-\bar{\mathbf{K}}\cdot\mathbf{D}_c^{(0)})^{-1}\cdot\bar{\mathbf{Q}} - \frac{1}{2}\mathbf{Tr}\ln(1-\mathbf{D}_c\cdot\bar{\mathbf{K}})\right] \\ &\cdot\exp\left[\frac{i}{2}\int A\cdot\bar{\mathbf{K}}\cdot(1-\mathbf{D}_c^{(0)}\cdot\bar{\mathbf{K}})^{-1}\cdot A\right. \\ &\left.+ i\int\bar{\mathbf{Q}}\cdot(1-\bar{\mathbf{K}}\cdot\mathbf{D}_c^{(0)})^{-1}\cdot A\right]. \end{aligned} \tag{9.31}$$

And one now sees that the kernel $\mathbf{D}_c^{(0)}\cdot(1-\bar{\mathbf{K}}\cdot\mathbf{D}_c^{(0)})^{-1}$ reduces to

$$\mathbf{D}_c^{(0)}\cdot(1-\bar{\mathbf{K}}\cdot\mathbf{D}_c^{(0)})^{-1} = \mathbf{D}_c^{(0)}\cdot[1-(\hat{\mathbf{K}}+(\mathbf{D}_c^{(0)})^{-1})\cdot\mathbf{D}_c^{(0)}]^{-1} = -\hat{\mathbf{K}}^{-1}, \tag{9.32}$$

where instead of (9.26), one now has

$$\hat{\mathbf{K}}^{ab}_{\mu\nu} = \tilde{\mathbf{K}}^{ab}_{\mu\nu} + gf^{abc}\chi^c_{\mu\nu}. \tag{9.33}$$

In the limit $A \to 0$, Equation (9.28) now yields

$$
\begin{aligned}
e^{\mathcal{D}_A}\mathcal{F}_1[A]\mathcal{F}_2[A] = \exp &\left[-\frac{i}{2}\int \bar{\mathbf{Q}}\cdot\hat{\mathbf{K}}^{-1}\cdot\bar{\mathbf{Q}} + \frac{1}{2}\mathbf{Tr}\ln\hat{\mathbf{K}} + \frac{1}{2}\mathbf{Tr}\ln(-\mathbf{D}_c^{(0)})\right] \\
\cdot \exp &\left[+\frac{i}{2}\int \frac{\delta}{\delta A'}\cdot\mathbf{D}_c^{(0)}\cdot\frac{\delta}{\delta A'}\right] \\
\cdot \exp &\left[\frac{i}{2}\int \frac{\delta}{\delta A'}\cdot\hat{\mathbf{K}}^{-1}\cdot\frac{\delta}{\delta A'} - \int \bar{\mathbf{Q}}\cdot\hat{\mathbf{K}}^{-1}\cdot\frac{\delta}{\delta A'}\right] \\
\cdot\, &(e^{\mathcal{D}_A}\ \mathcal{F}_2[A']).
\end{aligned} \tag{9.34}
$$

Now observe that the exponential term on the second line of (9.34) is exactly $\exp\{-\mathcal{D}_{A'}\}$ and serves to remove $\exp\{\mathcal{D}_{A'}\}$ of the operation $(\exp\{\mathcal{D}_{A'}\}\cdot \mathcal{F}_2[A'])$. With the exception of an irrelevant $\exp[\mathbf{Tr}\ln(-\mathbf{D}_c^{(0)})]$ factor, to be absorbed into an overall normalization, what remains to all orders of coupling for every such process is therefore the generic structure

$$
\begin{aligned}
e^{\mathcal{D}_A}\mathcal{F}_1[A]\mathcal{F}_2[A] = \mathcal{N}\exp &\left[-\frac{i}{2}\int \bar{\mathbf{Q}}\cdot\hat{\mathbf{K}}^{-1}\cdot\bar{\mathbf{Q}} + \frac{1}{2}\mathbf{Tr}\ln\hat{\mathbf{K}}\right] \\
\cdot \exp &\left[\frac{i}{2}\int \frac{\delta}{\delta A}\cdot\hat{\mathbf{K}}^{-1}\cdot\frac{\delta}{\delta A} - \int \bar{\mathbf{Q}}\cdot\hat{\mathbf{K}}^{-1}\cdot\frac{\delta}{\delta A}\right] \\
\cdot \exp &(\mathbf{L}[A]),
\end{aligned} \tag{9.35}
$$

in which the now-useless prime of A' has been suppressed. From (9.35) one may now draw the following conclusions:

(1) Nothing in Equation (9.35) refers to $\mathbf{D}_c^{(0)}$, which means that gauge invariance is here rigorously achieved as a matter of gauge independence. Such invariance cannot be more manifest.
(2) As expressed in Equation (9.35), each linkage operation upon $\exp\{\mathbf{L}[A]\}$ now consists of the exchange of a full "bundle" of gluons, represented by a factor of $\hat{\mathbf{K}}^{-1}$, while all of the cubic and quartic gluon interactions are conveniently incorporated into the remaining Halpern functional integral. This is quite different, in structure and interpretation, from linkages involving the corresponding $\mathbf{L}[A]$ of an Abelian theory, which entail the exchange of but a single virtual boson. It now becomes convenient to replace Feynman diagrams, containing different orders of gluon exchange, by "bundle diagrams", in which every graph depicts the non-perturbative exchange of all possible gluons between a pair of quark and/or anti-quark lines.

It should be noted that quite similar forms involving at least the part $(gf \cdot \chi)^{-1}$ of $\hat{\mathbf{K}}^{-1}$ were previously obtained in an instanton approximation to a gauge-dependent functional integral over gluon fluctuations. [Reinhardt, et al. (1993)][Hofmann (2006)] The present result, Equation (9.35), shows that such forms are an integral part of the exact QCD theory. This manifestly gauge invariant construction does not work for QED, where the simple rearrangement leading from Equations (9.12) to (9.14) cannot be implemented.

(3) A striking aspect of Equation (9.35) is that, because $\hat{\mathbf{K}} = \tilde{\mathbf{K}} + (gf \cdot \chi)$ and the $\tilde{\mathbf{K}}$ and $\tilde{\mathbf{Q}}$ coming from $\mathbf{L}[A]$ are all local functions, with non-zero matrix elements $\langle z|\tilde{\mathbf{K}}|z'\rangle = \tilde{\mathbf{K}}(z)\delta^{(4)}(z-z')$, the contributions of Equation (9.35) will depend on the Fradkin and Halpern variables in a specific but local way. This remarkable property will hereafter be called "Effective Locality" (EL), and one can now display the practical usefulness of this description in which relevant functional integrals can effectively be reduced to a few sets of ordinary integrals.

9.3 Effective Locality

It is worth emphasizing the locality aspects of the results in the previous section. Perhaps, the simplest example of EL is in the context of the simple, but nontrivial, quenched eikonal scattering model studied in [Hofmann (2006)], which contained the Halpern functional integral

$$\mathcal{N} \int d[\chi] e^{\frac{i}{4} \int \chi^2} \cdot [\det(gf \cdot \chi)]^{\frac{1}{2}} \cdot \exp\left[-\frac{i}{2} \int \bar{\mathbf{Q}} \cdot (gf \cdot \chi)^{-1} \cdot \bar{\mathbf{Q}}\right], \quad (9.36)$$

where the neglect of terms contributing to possible quark self-energies, in the limit of strong coupling, replaces the exponential factor of (9.36) by the argument

$$ig\varphi(b)\Omega_I^a[f \cdot \chi(w)]^{-1}\big|_{03}^{ab}\Omega_{II}^a. \quad (9.37)$$

The Fradkin u'-variables have been replaced by asymptotic 4-momenta by virtue of the simplifying eikonal approximation, and those momenta are automatically cancelled as will be made clear after (10.1) and (10.8). In (9.37), the color factors $\Omega_I^a(0)$, $\Omega_{II}^b(0)$ are peaked at $s_1 = 0$ and $s_2 = 0$, as will be noted shortly inconjunction with the function $\varphi(b)$, which depends on the collision's impact parameter b.

As a result of EL, the argument of the Halpern variable $\chi(w)$ of (9.37) was shown in [Fried, et al. (2012)] to be fixed at a specific value $w_0 = (0, \mathbf{y}_\perp, 0)$, where y denotes the CM spacetime coordinate of one of the scattering quarks or antiquarks. This is the only $\chi(w)$ that is relevant to the interaction. All of the other $\chi(w)$, for $w \neq w_0$ (and surrounded by a small volume of amount δ^4, as in the definition/construction of a functional integral) are simply removed from the problem along with their normalization factors, leaving a single, normalized functional integral over $d^n\chi(w_0)$.

Further, were the values of y to be subsequently changed, so that $w_0 \to w_1$, then all the $\chi(w)$, $w \neq w_1$, would become irrelevant, and with their normalization factors will cancel away. In effect, the spacetime index w is deprived of any physical meaning, and in this simple, quenched eikonal scattering, can be omitted. [When quenching is removed and different factors of $\mathbf{L}[A]$ are introduced, there can be more than one such $\int d^n\chi(w)$ to be performed; but the essential, simplifying effect of EL remains.] In brief, thanks to the EL property of this formulation, the Halpern functional integral can be reduced to a set of ordinary finite-dimensional integrals, which can be evaluated numerically, or approximated by a relevant physical approximation.

However, the final result for the amplitude will depend upon δ, and hence a non-trivial result will depend upon forming a connection between the vanishingly-small, mathematical δ, and a physically meaningful δ_{phi} consistent with experimental measurement. That this is both necessary and can be achieved from EL, can be seen as follows.

After EL has forced only a single $\chi(w_0)$ to appear in the interaction exponential, with all the other $w_\nu \neq w_0$ integrals trivially evaluated and combining with their normalization factors to give a net contribution of 1, we are left with the integral $N(\delta) \int d^n\chi(w_0)$ over the interaction part of the integral which, when evaluated, depends upon δ. Now define a rescaled variable $\bar{\chi}$ by the relation $\delta^4\chi^2 = \bar{\chi}^2\delta^4_{\text{phy}}$, so that the effective normalization and the new measure combine to give $N(\delta_{\text{phy}}) \int d^n\bar{\chi}$.

EL forces us to think in physical terms, and the only remaining question is how to choose δ_{phy} which quantity will now appear in the interaction integrand (multiplying $g\varphi(b)$); and we let Quantum Mechanics provide the answer: That contribution corresponding to a time separation should be chosen as $\frac{1}{E}$, that corresponding to a high energy CM longitudinal coordinate should be $\frac{1}{p_L} \simeq \frac{1}{E}$, while that corresponding to each of the transverse coordinates should be $\frac{1}{\mu}$, where μ is on the order of the pion mass; and hence $\delta^4_{\text{phy}} = (\frac{1}{\mu E})^2$.

To paraphrase Schwinger, just as renormalization is needed, in principle, to make the connection between interacting fields and asymptotic Abelian fields, so an effective renormalization is required to pass from the non-Abelian, gluon fields - whose quanta do not have asymptotic states and do not appear as part of the non-perturbative amplitude - to a particle description in which their multiple exchanges do provide a full asymptotic description. It is the Halpern field variables representing the GBs which are here effectively renormalized, and that renormalization is defined by the replacement of the mathematical δ by the physical δ_{phy}. We subsequently suppress the subscript of δ_{phy}, and the bar notation of $\bar{\chi}$.

Non-perturbative renormalization of the quark field is a separate matter, and turns out to be given in terms of GBs connecting closed quark loops to an individual quark line, as well as to each other. That investigation is presently in its earliest stages.

Chapter 10

QCD Transverse Fluctuations

10.1 Introduction

A basic distinction between QCD and other theories, in particular QED, must now be made, following from the results of Chapter 9 and is central to all applications to follow.

As noted in the QCD Introduction, this distinction occurs because the quanta of all (Abelian) quantized fields may be expected to obey standard quantum-mechanical properties, such as perfect position dependence at the cost of unknown momenta, and vice-versa. But this is impossible for quarks since they always appear asymptotically in bound states; their transverse positions or momenta can never, in principle, be exactly measured. Neglect of this distinction produces absurdities in the exact evaluation of all QCD amplitudes.

A phenomenological change in the basic QCD Lagrangian will accordingly be proposed, such that a probability amplitude of transverse fluctuations is automatically contained in the new Lagrangian, which eventually leads to those potentials essential to quark binding into hadrons, and hadron binding into nuclei. Then, all absurdities in estimates of QCD amplitudes are removed, enabling one to analytically calculate the effective potentials that produce quark binding as well as nucleon scattering and binding potentials.

Before proceeding with this phenomenological change, it is useful to describe the type of absurdity which will occur. As a typical and important part of a 4-point fermionic function, one finds an exponential factor of

$$
\begin{aligned}
&+ \frac{i}{2} g \int d^4w \int_0^s ds_1 \int_0^s ds_2 u'_\mu(s_1)\bar{u}'_\nu(s_2) \\
&\times \Omega^a(s_1)\bar{\Omega}^b(s_2)(f \cdot \chi(w))^{-1}|^{ab}_{\mu\nu} \\
&\times \delta^{(4)}(w - y_1 + u(s_1))\delta^{(4)}(w - y_2 + \bar{u}(s_2)). \qquad (10.1)
\end{aligned}
$$

This expression, corresponding to the interaction of particles 1 and 2 is obtained in the approximation of quenching and by neglecting quark spins. We emphasize that the full, exact expression displays exactly the same forms as the one under consideration, which is why the point can be made using this simplified example.

In (10.1), u_μ, Ω^a_1 and s_1 are variables associated with $\mathbf{G}^I_c(x_1, y_1|A)$, whereas $\bar{u}_\nu$, $\bar{\Omega}^b_I I$ and s_2 refer to $\mathbf{G}^{II}_c(x_2, y_2|A)$. The last line of (10.1) may be written as

$$
\delta^{(4)}(w - y_1 + u(s_1))\delta^{(4)}(y_1 - y_2 + \bar{u}(s_2) - u(s_1)) \qquad (10.2)
$$

and one sees that, as a consequence of EL, this interaction is peaked at $w_0 = y_1 - u(s_1)$. This means that all the other $w \neq w_0$ are irrelevant to the interaction, and, as discussed above, are removed along with their normalization factors, leaving dependence only upon an integration at one single point w_0, denoted by $\int d^n\chi(w_0)$.

The point central to the argument of this section has now being reached, as one evaluates the support of the second delta-function (10.2), which may be expressed as the product of delta-functions, in time, longitudinal and transverse coordinates,

$$
\begin{aligned}
&\delta(y_{10} - y_{20} + \bar{u}_0(s_2) - u_0(s_1)) \\
&\quad \times \delta(y_{1L} - y_{2L} + \bar{u}_L(s_2) - u_L(s_1)) \\
&\quad \times \delta^{(2)}(\mathbf{y}_{1\perp} - \mathbf{y}_{2\perp} + \bar{\mathbf{u}}_\perp(s_2) - \mathbf{u}_\perp(s_1)). \qquad (10.3)
\end{aligned}
$$

In the CM of quark 1 and quark 2, one can choose the origin of each time coordinate as the time of their closest approach, and then the time difference $y_{10} - y_{20}$ is always zero. If the Q_1 and Q_2 are scattering, then $y_{1L} + y_{2L} = 0$, since their longitudinal projections are in opposite directions; alternatively, if the Q_1 and Q_2 are bound together, then $y_{1L} = y_{2L}$, and their difference vanishes. Either choice makes no difference at all to the following analysis, and so we adopt the simplest, second possibility, $y_{1L} - y_{2L} = 0$.

Consider the time-coordinate delta-function, $\delta(\bar{u}_0(s_2) - u_0(s_1))$, which can have a zero argument whenever $\bar{u}_0(s_2)$ and $u_0(s_1)$ coincide. Assume this happens at a set of points s_l, so that

$$\begin{aligned}\delta(\bar{u}_0(s_2) - u_0(s_1)) &= \sum_l \delta(\bar{u}_0(s_l) - u_0(s_1) + (s_2 - s_l) \cdot \bar{u}_0'(s_l) + \cdots) \\ &= \sum_l \frac{1}{|\bar{u}_0'(s_l)|} \delta(s_2 - s_l)\Big|_{\bar{u}_0(s_l) = u_0(s_1)}. \end{aligned} \tag{10.4}$$

In a similar way, the longitudinal delta-function may be evaluated as

$$\sum_m \frac{1}{|u_L'(s_m)|} \delta(s_1 - s_m)\Big|_{u_L(s_m) = \bar{u}_L(s_2) \to \bar{u}_L(s_l)} \tag{10.5}$$

and their product as

$$\sum_{l,m} \frac{1}{|u_L'(s_m)|} \frac{1}{|\bar{u}_0'(s_l)|} \delta(s_1 - s_m)\delta(s_2 - s_l), \tag{10.6}$$

under the restrictions $u_0(s_m) = \bar{u}_0(s_l)$ and $u_L(s_m) = \bar{u}_L(s_l)$. Now, u_o and $\bar{u}_0$, and u_L and $\bar{u}_L$ are continuous but otherwise completely arbitrary functions: The probability that the intersections of $u_0(s_1)$ with $\bar{u}_0(s_2)$ and of $u_L(s_1)$ with $\bar{u}_L(s_2)$ occur at exactly the same points is therefore arbitrarily small. The only place where all four of these continuous functions have the same value is at $s_1 = s_2 = 0$, where, by definition of these functions, $u_\mu(0) = \bar{u}_\mu(0) = 0$; and hence, this pair of delta-functions collapse to the simple product,

$$\delta(\bar{u}_0(s_2) - u_0(s_1))\delta(\bar{u}_L(s_2) - u_L(s_3)) \to \frac{\delta(s_1)\delta(s_2)}{|u_L'(0)||\bar{u}_0'(0)|}. \tag{10.7}$$

By definition one has

$$u_\mu(s) = \int_0^s ds'\ u_\mu'(s'), \qquad u_\mu(0) = 0 \tag{10.8}$$

and then it is the remaining, transverse delta-function of (10.3) which is now most relevant, the term

$$\delta^{(2)}(\mathbf{y}_{1\perp} - \mathbf{y}_{2\perp}) = \delta^{(2)}(\mathbf{b}), \tag{10.9}$$

where $\mathbf{b}$ denotes the impact parameter, or transverse distance between the two scattering particles. This $\delta^{(2)}(\mathbf{b})$ appears as a multiplicative factor in the exponential of (9.35) and the question arises as to what meaning it can be assigned. Depending on its argument, a delta-function is either zero or infinite. In the first case this means that there is no interaction, while the second case means that at $\mathbf{b} = 0$, one has an infinite phase factor, suggestive of hard disc scattering.

The relevant question is therefore why such a delta-function $\delta^{(2)}(\mathbf{b})$ appears at all. The answer is that the assumption has earlier been made, in the conventional Abelian way, that the quark and/or anti-quark may be treated as ordinary particles, whereas asymptotic quarks exist only in bound states: Their transverse coordinates cannot, in principle, be specified, and there is therefore no reason to retain the conventional (Abelian) practice in which such measurement is assumed possible. This is the interpretation that will be adopted here, taking the $\delta^{(2)}(\mathbf{b})$ outcome as a serious warning that some form of input quark transverse fluctuation is necessary.

Why does this happens in QCD? Because QCD possesses EL, which conventional Abelian theories do not. The latter display sums over interconnected propagators, which provide a certain vagueness of position, whereas in the exact non-perturbative QCD, as described above, one finds the sharp determination of delta-functions corresponding to the EL property, and transverse imprecision must therefore be introduced separately, as a fundamental input to the theory.

In [Fried, et al. (2010)] it was suggested that this difficulty might be treated in an *ad hoc* phenomenological way, by replacing $\delta^{(2)}(\mathbf{b})$ by the smoothly varying, effective Gaussian

$$\varphi(\mathbf{b}) = (2\pi)^{-2} \int d^2\mathbf{k} e^{i\mathbf{k}\cdot\mathbf{b} - \frac{k^2}{4\mu^2}},$$

where μ is a mass parameter on the order of the $Q - \bar{Q}$ bound state (which we shall call a "model pion"), although we were able to obtain the conclusions of that paper without specifying the precise form of $\varphi(\mathbf{b})$. Here we face this question directly, by first developing a formalism in which transverse quark coordinates cannot be specified, and then showing how this formalism removes all such absurdities, such as that of the exponential factor of $\delta^{(2)}(\mathbf{b})$ above. But it must be emphasized that our prescription is phenomenological, for there remains to be shown how such an approach could be derived from a more fundamental, operator-field version of QCD, in which transverse fluctuations would occur automatically.

10.2 A Phenomenological Expression of Transverse Imprecision

Perhaps the simplest way of introducing transverse fluctuations is to average that part of the QCD Lagrangian dealing with the quark-gluon interaction, so that the transverse position of the color-charge current operator $\bar{\psi}\gamma_\mu\tau^a\psi(x)$ should be averaged over a small range by means of an initially unspecified distribution. One can also demand the same imprecision for the vector current $\bar{\psi}\gamma_\mu\psi(x)$ and scalar density $\bar{\psi}\psi(x)$, but these extra requirements seem to complicate the presentation, to no real advantage, and will not be considered here.

It should be emphasized that the following method is perhaps the simplest phenomenological way of introducing a measure of such transverse imprecision/fluctuation; and that there are quite possibly other, more profound methods of obtaining the same objective. In fact, it seems most probable that a lack of anti-commutation of quark operator fields at equal times, but at differing transverse coordinates, could be the basic point requiring attention. In the interests of simplicity and clarIty, we ask the readers' indulgence for postponing that particular investigation.

Instead of the conventional quark-gluon contribution to the Lagrangian density,

$$\mathcal{L}_{\text{QC}} = -\bar{\psi}[m + \gamma_\mu(\partial_\mu - igA^a_\mu\tau^a)]\psi, \tag{10.10}$$

in which all field operators occur at the same spacetime point, and for which gauge invariance under the standard QCD gauge transformations is obvious, we now adopt a local — in time and longitudinal position — but non-local in its transverse coordinates replacement,

$$\mathcal{L}'_{\text{QC}} = ig \int d^2\mathbf{x}'_\perp \; a(\mathbf{x}_\perp - \mathbf{x}'_\perp)\bar{\psi}(x')\gamma_\mu A^a_\mu(x)\tau^a\psi(x'), \tag{10.11}$$

where the transverse imprecision function (TIF) $a(\mathbf{x}_\perp - \mathbf{x}'_\perp)$ is a real, symmetric function of its arguments, of significant value only for distances on the order of the inverse of the pion mass; $x'_\mu = (x_0, \mathbf{x}'_\perp, x_L)$ and $A^a_\mu(x)$ is left untouched. In this formulation, rigorous local gauge-invariance is suppressed for the underlying quark fields, whose quanta have unmeasurable transverse positions, but the hadrons all constructed from these quanta will nevertheless be proper singlets under SU(3).

In fact, one may argue that required gauge invariance is maintained when quark properties are averaged over distances below which no measurement of their transverse properties can be physically performed. It would be most attractive, if and when a better formalism is invented, if strict gauge invariance could be maintained for all values of transverse coordinates; but that is not a physical requirement, but rather a mathematical nicety.

One notes that in the contribution of (10.11) to its part of the Action operator, $\int d^4x\mathcal{L}_{\mathrm{QC}}$, the $\mathbf{x}_\perp$ and $\mathbf{x}'_\perp$ coordinates can be interchanged, which yields an equivalent form in which every $A^a_\mu(x)$ of the original (10.10) is replaced by $\int d^2\mathbf{x}'_\perp a(\mathbf{x}_\perp - \mathbf{x}'_\perp)A^a_\mu(x')$. This interchange allows a very simple extraction of all such transverse imprecision, since both delta-functions of (10.1) will now be replaced by

$$\int d^2\,\mathbf{y}'_{1\perp}a(\mathbf{y}_{1\perp} - \mathbf{y}'_{1\perp}) \int d^2\mathbf{y}'_{2\perp}a(\mathbf{y}_{2\perp} - \mathbf{y}'_{2\perp}) \\ \times\, \delta^{(4)}(w - y'_1 + u(s_1))\delta^{(4)}(w - y'_2 + \bar{u}(s_2)). \tag{10.12}$$

One small complication of this procedure is that such "primed" transverse coordinates will now appear in the arguments of χ, e,g:, $\mathbf{w}_\perp \to \mathbf{y}'_\perp$, and the fixed position coordinate $\mathbf{w}_\perp$ must now itself be varied. But the difference $|\mathbf{y}_\perp - \mathbf{y}'_\perp|$ is effectively bounded by $1/\mu$, where μ appears in the definition of $\varphi(x)$ below, and it turns out that a negligible error is made when $\mathbf{w}_\perp$ is replaced by $\mathbf{y}_\perp$. More details are given in Appendix B of [Fried, et al. (2012)], where it is shown that this approximation is justified for the subsequent calculations of quark and nucleon bindings. And the argument for the irrelevance of changes in the $\mathbf{w}_\perp$ argument of $(f \cdot \chi(w))^{-1}$ can be extended to justify the replacement of $\mathbf{y}'_\perp$ by $\mathbf{y}_\perp$.

The first so-modified delta-function of (10.12) defines the argument w_1 of $\chi(w_1)$, and we again observe that the final output of the Halpern FI will be an ordinary integral $\int d^n\chi$, independent of the choice of w_1. The second delta-function of (10.12) now involves the a-dependence, generating in place of the $\delta^{(2)}(\mathbf{y}_{1\perp} - \mathbf{y}_{2\perp})$, the combination

$$\int d^2\,\mathbf{y}'_{1\perp} \int d^2\mathbf{y}'_{2\perp}a(\mathbf{y}_{1\perp} - \mathbf{y}'_{1\perp})a(\mathbf{y}_{2\perp} - \mathbf{y}'_{2\perp})\delta^{(2)}(\mathbf{y}'_{1\perp} - \mathbf{y}'_{2\perp}) \\ = \int d^2\mathbf{y}'_\perp a(\mathbf{y}_{1\perp} - \mathbf{y}'_\perp)a(\mathbf{y}_{2\perp} - \mathbf{y}'_\perp). \tag{10.13}$$

Inserting two-dimensional Fourier transforms of each $a(\Delta y)$, for example,

$$a(\mathbf{y}_{1\perp} - \mathbf{y}'_{\perp}) = \int \frac{d^2\mathbf{k}}{(2\pi)^2} e^{i\mathbf{k}\cdot(\mathbf{y}_{1\perp} - \mathbf{y}_{\perp})} \tilde{a}(\mathbf{k}_{\perp}), \tag{10.14}$$

the combination (10.13) becomes

$$\int \frac{d^2\mathbf{k}}{(2\pi)^2} \tilde{a}(\mathbf{k})\tilde{a}(-\mathbf{k}) e^{i\mathbf{k}\cdot(\mathbf{y}_{1\perp} - \mathbf{y}_{2\perp})}. \tag{10.15}$$

From its definition, a is real, and hence (10.15) becomes

$$\int \frac{d^2\mathbf{k}}{(2\pi)^2} e^{i\mathbf{k}\cdot\mathbf{b}} \left| \tilde{a}(\mathbf{k}) \right|^2 \equiv \varphi(\mathbf{b}), \tag{10.16}$$

which provides the definition of $\varphi(\mathbf{b})$. Note that while no restriction has been placed on the form of a other than that it is real and symmetric, φ turns out to be independent of the direction of $\mathbf{b}$, that is, $\varphi(\mathbf{b}) = \varphi(b)$. In this way, the improper $\delta^{(2)}(b)$ is automatically replaced by $\varphi(b)$.

10.3 Bundle Diagrams

In the above example of quark and/or antiquark scattering, where the infinite number of exchanged gluons appears to originate and end at a single spacetime point on a quark/antiquark line, *modulo* transverse imprecision it may be helpful to introduce the concept of an exchanged GB, as in Figure 10.1. Because of the four-dimensional delta function $\delta^{(4)}(y'_1 - y'_2 - u(s_1) + \bar{u}(s_2))$, arising from the product of the pair of delta functions of (10.12), and of the subsequent analysis which produces (10.13), the transverse separation $\mathbf{b} = \mathbf{y}_{1\perp} - \mathbf{y}_{2\perp}$ satisfies the probability distribution (10.16). The argument w in $(f \cdot \chi(w))^{-1}$ is given by $w = w_1 = y'_1 - u(s_1) = y'_1 \to y_1$, by virtue of (10.7) and, as explained above, the Halpern functional integral reduces to a set of ordinary integrals that are represented by the Bundle Diagram of Figure 10.1.

Here, therefore, it is understood that time and longitudinal coordinates of the end-points of the bundle are the same, whereas their transverse coordinates, measured vertically in the figure, are separated. Bundle diagrams are not Feynman diagrams, but offer perhaps a more efficient way of representing the sum over all of the Feynman graphs corresponding to such multiple gluon exchange.

A slightly more complicated expression describes GBs exchanged between any two of three quarks, as in Figure 10.2, where, because of EL, the

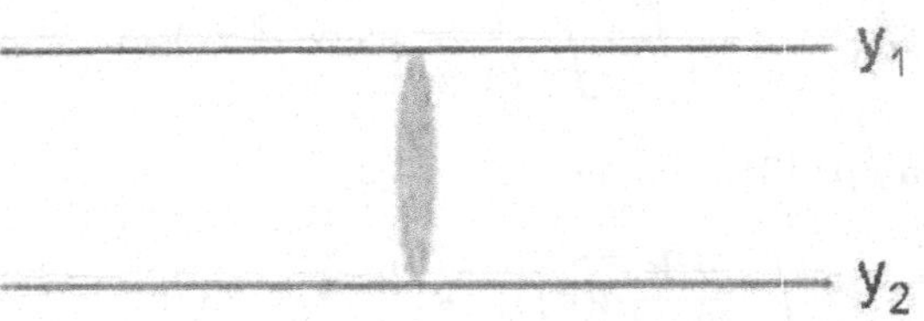

Fig. 10.1 A GB exchanged between quarks I and II.

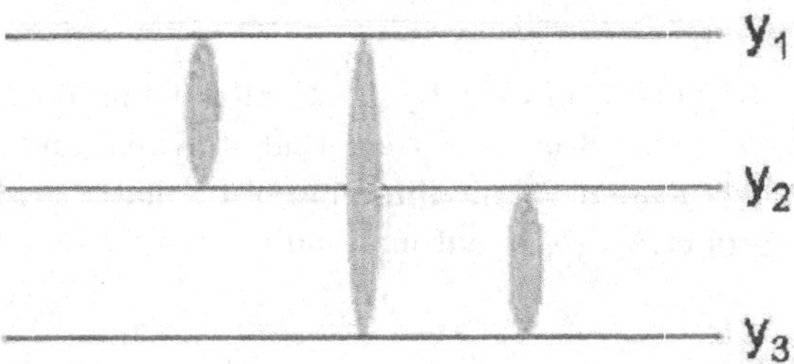

Fig. 10.2 GBs exchanged among three quarks.

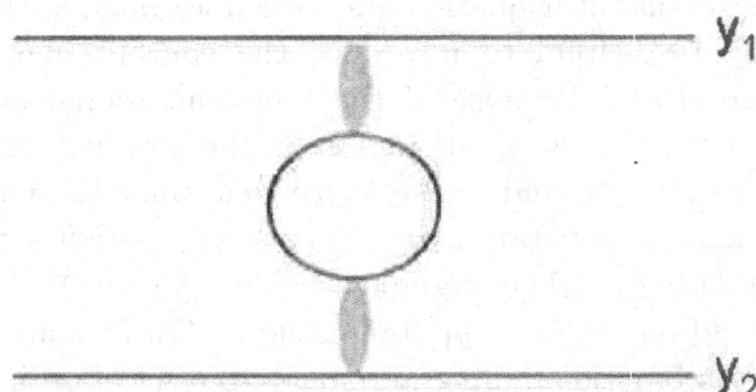

Fig. 10.3 GBs joining two quarks via a closed quark loop.

w-coordinates of each of the $(f \cdot \chi)^{-1}$ entering into the appropriate Halpern functional integral are the same.

The longitudinal and time-coordinates of the three quarks when forming a nucleon have the same values, although their transverse coordinates can be quite different.

In contrast, were a closed quark loop, corresponding to a simple relaxation of the quenched approximation, to appear between a pair of quarks, joined to each external quark line by the exchange of a GB, as in Figure 10.3, there will now be two distinct sets of ordinary Halpern integrals to be evaluated.

As will be seen in Chapter 12, when the quark lines are replaced by nucleons, the diagram of Figure 10.3 will provide us with the essential features of the nucleon-nucleon potential for separation lengths beyond $\frac{1}{m_\pi}$.

Chapter 11

Quark Binding Potential

11.1 Use of the Eikonal Approximation

An effective potential as the result of the exchange of multiple virtual gluons between Q's and/or $\bar{Q}$'s will now be derived. The word "effective" here represents the type of potential one would expect to find were the situation that of a scattering of conventional Abelian particles, moving at large relative velocities in their CM, as appropriate to an eikonal context. The word "Abelian" is relevant in this QCD situation, for the analysis of [Fried, et al. (2010)] suggests that for large impact parameters the scattering of such non-Abelian Q's and/or $\bar{Q}$'s is effectively coherent; only as the impact parameter decreases do progressively stronger color fluctuations appear and decrease the effective coupling, which then becomes the statement of non-perturbative asymptotic freedom in the variable conjugate to the impact parameter. For such scattering the eikonal model is, by far, the simplest method of extracting the relevant Physics.

We shall be interested in binding, or "restoring" potentials, effective potentials which prevent bound particles from remaining at relatively large separations. Since large impact parameters are relevant to large, three-dimensional distances, what is needed is, firstly, a statement of the relevant formulae of at large impact parameter; and, secondly, the corresponding eikonal function found in that limit. It is then a relatively simple matter to determine an effective potential which would produce the same eikonal.

A two-body eikonal scattering amplitude will have the form [Fried (1990)]

$$\tau(s,t) = \frac{is}{2m^2} \int d^2\mathbf{b} e^{i\mathbf{q}\cdot\mathbf{b}} [1 - e^{i\mathbf{X}(s,\mathbf{b})}], \tag{11.1}$$

where s and t denote the standard Mandelstam variables, $s = -(p_1 + p_2)^2, t = -(p_1 - p_1')^2 = \mathbf{q}^2$ in the CM of $Q\bar{Q}$, and where $\mathbf{X}(s,b)$ is the eikonal function. Following the arguments of [Fried, et al. (2010)]. One reaches

$$e^{i\mathbf{X}} = \mathcal{N} \int d^n\bar{\chi}_{30}[\det(gf\cdot\bar{\chi})^{-1}]^{\frac{1}{2}} \cdot e^{i\frac{\bar{\chi}^2}{4}} \cdot \exp[ig\varphi(b)\Omega_1^a \cdot (f\cdot\bar{\chi})^{-1}|_{30}^{ab} \cdot \Omega_{II}^a] \tag{11.2}$$

and the normalization is such that for $g\varphi(b) \to 0$, $\exp[i\mathbf{X}] \to 1$. Here, $n = N_c^2 - 1$, that is $n = 8$ for $N_c = 3$. All the integrals over

$$\int d^n\alpha_I \int d^n\alpha_{II} \int d^n\Omega_I \int d^n\Omega_{II} \tag{11.3}$$

are properly normalized, and they connect the $\Omega_I^a, \Omega_{II}^b$ dependence (11.2) with the $\lambda_I^a, \lambda_{II}^b$ Gell-Mann matrices placed between initial and final state vectors, which define the type of amplitude desired, $Q - Q$ scattering or $Q - \bar{Q}$ scattering, etc.

As in [Fried, et al. (2010)] the eight-dimensional integral over $\int d^n\bar{\chi}_{03}$ is arranged into the form of a radial integration over the magnitude of $\bar{\chi}_{03}$, now called R, and normalized angular integrations. Only the $0,3$ subscripts of χ appear in this simple analysis, but no such restriction is really necessary.

We suppress the latter, along with all normalized $\int d^n\Omega_I \int d^n\Omega_{II}$ dependence, and rewrite (11.2) in the form

$$e^{i\mathbf{X}(s,b)} = \mathcal{N}' \int_0^\infty dR\, R^3 e^{i\frac{R^2}{4} + i\frac{\langle g\rangle\varphi(b)}{R}}, \tag{11.4}$$

where the normalization of the R-integral is given by $\mathcal{N}'^{-1} = \int_0^\infty dR\, R^3 e^{i\frac{R^2}{4}}$, and where $\langle g\rangle$ is a shorthand denoting the effective coupling after all angular and color integrations and matrix elements have been calculated. In this calculation, only the linear dependence of $g\varphi$ is needed, while the non-vanishing matrix elements between singlet color states will define $Q - \bar{Q}$ scattering. The integer power of 3 in (11.4), of the term $R^3 = R^{7-4}$, arises from the $\frac{1}{2}\mathbf{TR}\ln\hat{K}$ factor of (9.35), contributing R^{-4} to (11.2). As noted above, we wish to compare this eikonal function at fairly large impact parameter with that corresponding to the scattering of two, interacting particles, and then infer what form of effective potential corresponds to that eikonal. We therefore assume the usual formula [Fried (1990)] relating an eikonal to a specified potential,

$$\mathbf{X}(s,b) = -\int_{-\infty}^{\infty} dz V(\mathbf{b} + z\hat{p}_L) = \gamma(x)\mathbf{X}(b), \tag{11.5}$$

where $\gamma(s)$ depends upon the CM energy of the scattering particles (and the nature of their interaction), and $\hat{p}_L$ is a unit vector in the direction of longitudinal motion. Equation (11.5) is true for nonrelativistic and relativistic scattering, in potential theory and in field theory. For large b, our eikonal turns out to be weakly dependent on s, and the b-independent contributions do not contribute to $V(r)$.

To obtain $V(r)$ from a given $\mathbf{X}(b)$ is then simply a matter of reversing the usual calculation: One computes the Fourier transform $\tilde{\mathbf{X}}(k_\perp)$ of $\mathbf{X}(b)$, and then extends $k_\perp$ to $|\mathbf{k}| = [\mathbf{k}_\perp^2 + k_L^2]^{\frac{1}{2}}$. The three-dimensional Fourier transform of $\tilde{\mathbf{X}}(k)$ is then proportional to $V(r)$. Following (10.16) we assume that $\mathbf{X}(b)$, which depends on the form of $\varphi(b)$, is a function of $\mathbf{b}^2$; and although we begin by asking for the form of $\varphi(b)$ for large b, it can be shown that the small-b contributions to the Fourier transforms are unimportant for the large-r behavior of $V(r)$.

One further point requires consideration, for the above remarks are valid when the process involves only the scattering of two particles. But when initial energies are high enough such that inelastic particle production occurs, then the potential corresponding to such production must have a negative imaginary component (so that the corresponding S-matrix, $\exp[-i\mathcal{H}t]$, will have a decaying time dependence), thereby conserving probability as a diminution of the final, two-particle state when production becomes possible. This is also true when the two initial particles have the possibility of binding, and forming a state which was not initially present; the probability of the final scattering state must diminish. In other words, the eikonal function one expects to find may always be characterized by a complex potential of form $V_R - iV_I$, where in the present case, V_I is that potential which can bind the Q and $\bar{Q}$ into a pion: In a scattering calculation, that potential must therefore appear with a multiplicative factor of $-i$ (symbolically, $i\mathbf{X} \to i(V_R - iV_I)$, and for this calculation of binding, $i\mathbf{X} \to V_I = V_B$).

Returning to (11.1), it is assumed that $\mathbf{q} \neq 0$, so that the "1" term of the integrand contributes a $\delta^{(2)}(q)$ that can be discarded. This amounts to a physical assumption: Q and $\bar{Q}$ are not bound rigidly, with an unchanging impact parameter; rather, there is a permanent, if small, "in-and-out" transverse motion, which represents the bound state, and hence $\mathbf{q} \neq 0$. Then

$$\mathcal{T}(s,t) \sim \int d^2b\, e^{i\mathbf{q}\cdot\mathbf{b}} e^{i\mathbf{X}(b)}, \tag{11.6}$$

where $\exp[i\chi(b)]$ is given by (11.4). In [Fried, et al. (2010)] it was assumed that $\varphi(b)$ could be a smooth function which vanishes for large b, such as $\sim \exp[-(\mu b)^2]$, but it turns out that an appropriate choice is

$$\varphi(b) = \varphi(0) e^{-(\mu b)^{2+\xi}} \tag{11.7}$$

where ξ is real, positive and small. For any such choice of ξ, $\varphi(b)$ becomes small as $\mu b \gg 1$.

For large b it is then sensible to expand the $\exp[i\langle g\rangle\varphi(b)/R]$ term of (11.4), retaining only the linear $\langle g\rangle\varphi$ dependence. This gives, in place of (11.4),

$$\begin{aligned} e^{i\mathbf{X}(b)} &\simeq \mathcal{N}' \int_0^\infty dR\, R^3 e^{i\frac{R^2}{4}} \left[1 + i\langle g\rangle \frac{\varphi(b)}{R} + \cdots\right] \\ &= 1 + ik\mathcal{N}'\langle g\rangle\varphi(b) + \cdots, \end{aligned} \tag{11.8}$$

where

$$k\mathcal{N}' = \int_0^\infty dR\, R^3 ei\frac{R^2}{4} = 2\sqrt{\pi}(-i)^{\frac{3}{2}} \tag{11.9}$$

and $\mathcal{N}' = -\frac{1}{8}$. Remembering that both sides of (11.8)) are to be integrated over $\int d^2\mathbf{b}e^{i\mathbf{q}\cdot b}$, for $\mathbf{q} \neq 0$ we can proceed to the identification

$$e^{i\mathbf{X}(b)} = ik\mathcal{N}'\langle q\rangle\varphi(b) \tag{11.10}$$

or

$$i\mathbf{X}(b) = \ln[\varphi(b)] + \cdots, \tag{11.11}$$

where $\ln\varphi$ is large and φ is (effectively) small, and where the dots will be commented on shortly.

Choosing $\varphi(b) = \varphi(0)e^{-(\mu b)^{2+\xi}}$, one obtains

$$i\mathbf{X}(b) = -(\mu b)^{2+\xi} + \cdots. \tag{11.12}$$

It is then convenient to use the integrals

$$\int_0^\infty dx\ x^\nu\ J_0(ax) = 2^\nu a^{-1-\nu}\Gamma\Big(\frac{1}{2}+\frac{\nu}{2}\Big)\Big/\Gamma\Big(\frac{1}{2}-\frac{\nu}{2}\Big), \quad \nu < \frac{1}{2} \tag{11.13}$$

and

$$\int_0^\infty dx\ x^{\nu-1}\sin(x) = \Gamma(\nu)\sin\Big(\frac{\nu\pi}{2}\Big), \quad |\mathbf{Re}(\nu)| < 1, \tag{11.14}$$

in which, except for obvious poles, the Gamma functions are analytic in ν and can be continued to the needed values. With

$$\begin{aligned} i\tilde{\mathbf{X}}(k_\perp) &= -\int d^2b(\mu b)^{2+\xi}e^{i\mathbf{k}_\perp\cdot\mathbf{b}} \\ &= -(2\pi)\int_0^\infty db\ J_0(k_\perp b)\mu^{2+\xi}b^{3+\xi} \end{aligned} \tag{11.15}$$

and the use of the doubling formula for Gamma functions, working everything through, one finds

$$V_B(r) = -\frac{2^{3+\xi}}{\pi}\mu^{2+\xi}r^{1+\xi}\frac{\Gamma(2+\frac{\xi}{2})}{\Gamma(-1-\frac{\xi}{2})}\Gamma(-2-\xi)\sin\Big(\frac{\pi\xi}{2}\Big), \tag{11.16}$$

and for small enough ξ, the confining potential

$$V_B(r) \simeq \xi\mu(\mu r)^{1+\xi}, \tag{11.17}$$

which can be compared to the results of several machine groups. [Nambu (1979)][Luscher, et al. (1980)][Luscher (1981)]. It is remarkable that the choice (11.7) can be viewed as part of a Levy-flight probability distribution that generalizes Gaussian ones in consistency with the famous *Central Limit Theorem* of statistical physics. It is interesting to note that at a physical level, such a distribution[1] addresses the issue of transverse quark motions, describing them in a way that can seem quite relevant to a confined context.

There remains the question of what values should be assigned to ξ, and to the unknown parameter μ/m_Q. These issues will be dealt with in the next section. Incidentally, one may observe that in (11.12), additive constants to $\ln[\varphi(b)]$ bring no contribution to $V_B(r)$.

[1]For example, see *http://en.wikipedia.org/wiki/stable distribution*

To calculate the corresponding effective restoring potential when one of the three quarks contributing to the QQQ bound state is separated from the other two, or, more simply, when all three quarks are forced to separate from each other, one may refer to Equation (22) of [Fried, et al. (1983)] giving the eikonal amplitude corresponding to Coulomb three-particle scattering. Here, the specifically Coulomb parts of this amplitude may be replaced by the single Halpern integral which connects the three quarks to each other, as it connected the Q and $\bar{Q}$ to each other in the calculation above. If particle 2 of this formula enters the scattering with zero transverse momentum in the rest frame of particles 1 and 3, the amplitude simplifies to

$$\mathcal{T}_3^{eik} \sim \int d^2\vec{b}_{12} \int d^2\vec{b}_{32} e^{i\vec{q}_3\cdot\vec{b}_{32}+i\vec{q}_1\cdot\vec{b}_{12}} \left[\Phi(\vec{b}_{12},\vec{b}_{32},\vec{b}_{13}) + \Psi_3\right], \quad (11.18)$$

where $\vec{b}_{13} = \vec{b}_{12} - \vec{b}_{32}$, irrelevant kinematic factors multiplying these integrals have been suppressed, and Ψ_3 denotes the combination

$$-1+\left[1-\Phi(\vec{b}_{12},\infty,\infty)\right]+\left[1-\Phi(\infty,\vec{b}_{32},\infty)\right]+\left[1-\Phi(\infty,\infty,\vec{b}_{13})\right], \quad (11.19)$$

so that $\mathcal{T}_3^{eik}$ is a completely "connected" amplitude.

Writing Φ as $\exp[i\mathbf{X}(b_{12},b_{32})]$, its defining integral may be rewritten as

$$\mathcal{N}' \int_0^\infty dR\, R^3 \exp\left[i\frac{R^2}{4} + i[\langle g\rangle_{12}\varphi(b_{12}) + \langle g\rangle_{32}\varphi(b_{32}) + \langle g\rangle_{13}\varphi(b_{13})]\right], \quad (11.20)$$

and one may ask for the form (11.19) takes when one or more of the b_{ij} becomes large.

When any one of the b_{ij} becomes large, $\varphi(b_{ij})$ becomes small, and its exponential term may be expanded. The simplest situation is when they all become large, so that an expansion of (11.18) is relevant. The first non-zero and leading term in such an expansion which contains the φ dependence, and hence the b_{ij}-dependence, of all three quarks contributing to the singlet QQQ state is that term for which the expansion yields

$$1 + k\mathcal{N}'[\langle g\rangle_{12}\varphi(b_{12}) + \langle g\rangle_{32}\varphi(b_{32}) + \langle g\rangle_{13}\varphi(b_{13})] + \cdots, \quad (11.21)$$

so that, suppressing all normalized angular and color integrations, for $\vec{q}_1 \neq 0, \vec{q}_3 \neq 0$,

$$e^{i\mathbf{X}(b_{12},b_{32})} \simeq k\mathcal{N}'[\langle g\rangle_{12}\varphi(b_{12}) + \langle g\rangle_{32}\varphi(b_{32}) + \langle g\rangle_{13}\varphi(b_{13})] + \cdots. \quad (11.22)$$

If particles 1 and 3 are now additionally separated, keeping the distances b_{12} and b_{32} essentially fixed, then

$$\begin{aligned} e^{i\mathbf{X}(b_{12},b_{32})} &\simeq k\mathcal{N}'\langle g\rangle_{13}\varphi(b_{13}) + \cdots\,, \\ i\mathbf{X}(b_{12},b_{32}) &\simeq \ln[\varphi(b_{13}] + \cdots\,, \end{aligned} \tag{11.23}$$

and we are effectively back in the "pion" situation, where the large-impact parameter eikonal of $Q-\bar{Q}$ scattering was calculated. Clearly, (11.23) suggests that when any two of the three quarks which contribute to the singlet "nucleon" are well separated, there results a restoring potential $V(r_{ij})$ of the same form as that of (11.17). There are, of course, corrections to this confining potential, those which are obvious from the approximations used above, as well as those corresponding to spin and angular momentum, which have been completely neglected.

11.2 Estimation of the "Model Pion" Mass

This Section contains an estimate of the effects of the above binding potential for the simplest case of the "pion", a ground state of the $Q-\bar{Q}$ system calculated using the simplest minimization technique [Balibar (1989)] as well as a second minimization with respect to a ratio of terms introduced in this analysis.

In this simplest, non-relativistic estimation, the Hamiltonian of this system is given by

$$\mathcal{H} = 2m + \frac{1}{m}p^2 + V(r), \tag{11.24}$$

where $1/m$ denotes the inverse of the reduced mass of this equal-mass quark system. The ground-state energy is then given by the replacement of p by $1/r$, and the subsequent minimization of the eigenvalue of this Hamiltonian with respect to r. Assuming $\xi \ll 1$, one obtains

$$\mu r = \left(\frac{2}{\xi}\frac{\mu}{m}\right)^{1/3} \tag{11.25}$$

and

$$E_0 = 2m + \frac{3}{2}\xi\mu\left(\frac{2}{\xi}\frac{\mu}{m}\right)^{1/3}. \tag{11.26}$$

Here, μ and ξ are the two parameters required by this analysis in order to have a non-zero binding potential; and while their existence has been introduced as a necessary assumption for the interacting $Q-\bar{Q}$ system, there is another unknown quantity, the quark mass m, which must play an important role. If we take the position that μ and ξ will appear as a result of a more fundamental QFT (in which the transverse momenta or position coordinates of the quanta of the QCD fields can never be measured with precision), then we are free to consider which values of m, or of the ratio μ/m, can lead to the lowest value of E. In so doing, one is asking if there might exist a dynamical reason for the relatively small value of the pion mass, in this approximation to the actual pion; that is, whether approximate chiral symmetry has a dynamical basis.

Following this approach and using the small-ξ limit of (11.26),

$$E_0 = \mu\left[2\left(\frac{\mu}{m}\right)^{-1} + 3\left(\frac{\xi}{2}\right)^{2/3}\left(\frac{\mu}{m}\right)^{1/3}\right]. \tag{11.27}$$

Then, upon variation with respect to $x = \mu/m$, the function $E_o(x)$ will have an extremum at $x_0 = 2^{3/4}\left(\frac{2}{\xi}\right)^{1/2}$; and since

$$\left.\frac{\partial^2 E_0(x)}{\partial x^2}\right|_{x_0} = \mu 2^{-5/4}\left(\frac{\xi}{2}\right)^{3/2}\left(2-\frac{2}{3}\right) > 0, \tag{11.28}$$

that extremum is a minimum. Substituting the value of $\mu/m = 2^{3/4}\left(\frac{2}{\xi}\right)^{1/2}$ into (11.27), yields

$$E_0 = \mu\xi^{1/2}\, 2^{-1/4}[1+3], \tag{11.29}$$

from which one infers that there is three times as much energy in the gluon field and $Q-\bar{Q}$-kinetic energies as in the quark rest masses. Intuitively, one expects that $E \sim m_\pi \sim \mu$, which suggests that $\xi \sim \sqrt{2}/16$. And finally, one then has an estimate of the quark mass, in terms of ξ and μ.

Of course, this double-minimization calculation of E_0 cannot be exactly identified with the precise pion mass, because of the approximations made above, as well as the omission of more complicated singlet terms, such as the contributions coming from $QQ\bar{Q}\bar{Q}$ terms. But the Physics seems to be reasonably correct. Note also that we are dealing with "textbook" QCD, containing but one type of quark and its complement of massless, SU(3) gluons; flavors and electro-weak corrections can be added, as desired. A similar minimization analysis can be made for the nucleon ground state, but there then are other degrees of freedom which should be included, and treated properly by a serious attempt at a solution of such a three-body, relativistic problem. This the author must leave to others whose numerical abilities far exceed his own.

It should be emphasized that a comparison of experimental pion and nucleon masses with the minimal-energy values of these (and improved) model $Q\bar{Q}$ and QQQ calculations serves, in principle, to determine the ξ, μ parameters, which are contained within the transversally-modified Lagrangian. If either parameter is set equal to zero, there is no quark binding. It may be somewhat unexpected that there are two such relevant parameters, a mass scale and a dimensionless quantity defining, in part, the needed transverse fluctuations; but a glance at (11.17) shows that both are necessary. With the simplifying approximations of this Chapter, we understand that $\xi \sim 0(1/10)$, and $\mu \sim 0(m_\pi)$.

Finally, it should be mentioned that the quenched, eikonal amplitude for GB exchange between quarks and/or antiquarks can be calculated without recourse to the simplifications of this Chapter. The result of this exact treatment — whose simplest, relevant approximation yields the results found above — is an amplitude given in terms of Meijer G-functions, which contain dependence upon kinematic variables appropriate to a scattering process. Details may be found in [Grandou, et al. (2013)].

Chapter 12

Nucleon Scattering and Binding

12.1 Introduction

In this Chapter, we begin with the concept of three bound quarks scattering against another triad of three bound quarks, with their full, non-perturbative exchanges of gluons between all the quarks taking place. We assume that those triads that are initially bound remain bound at all times, which carries the implication that the multiple gluons exchanged between these nucleons do not change the overall color-singlet nature of each nucleon. We neglect all electroweak interactions; and to further simplify the analysis assume that this nucleon scattering takes place at high relative velocities, so that a simplifying eikonal description of the scattering may be used. We further simplify the analysis by neglecting spin effects — which could be inserted if desired — and aim for a simple, qualitative picture of how forces between nucleons can arise, starting from the basic fundamentals of realistic QCD.

In one sense, however, our eikonal model must be made more complicated than those quenched models used previously, for it turns out that one must here retain at least the simplest effects of the closed-quark-loop, or vacuum functional, $\mathbf{L}[A]$. The Physics underlying this requirement follows because the forces which arise between quarks due to the multiple exchange of gluons are strong, tending to bind, for impact parameter b separations on the order of $1/\mu$, but the amplitude for this process falls off rapidly with increasing $b > 1/\mu$. (At distances $b \ll 1/\mu$ large color fluctuations tend to reduce the value of any amplitude, which corresponds to non-perturbative asymptotic freedom). How can nucleons, whose internal structures are defined at distances $b \sim 1/\mu$, then feel strong forces at separations $b > 1/\mu$?

The answer is that vacuum loops, defined by $\mathbf{L}[A]$, can stretch in the transverse directions, and can serve to transmit the multiple gluon interactions across larger values of impact parameter; a gluon "bundle" from one nucleon attaches itself to one point on the loop, while another bundle of gluons passes from a second point on the loop (at a significant transverse distance from the first) to the other nucleon. Although this passage of momentum via a closed vacuum loop changes the interaction somewhat, an essential "short-range" interaction is produced at distances larger than would be possible by quark-to-quark passage alone. And if, as this loop is stretched in a transverse direction, one transverse side of the loop corresponds to a quark and the other to an antiquark, one has the image of an effective pion being exchanged between scattering nucleons.

For reasons of simplicity, we shall replace the $\mathbf{G}_c[A]$ of each quark in a nucleon, and the single $\mathcal{G}_c[A]$ which models that nucleon, by its high-energy eikonal limit, and then connect the gluon bundles emitted by each nucleon to two, and only two points on a single loop. More complicated loop structures are certainly possible, and should be investigated, but this is the simplest representation of "effective pion exchange between nucleons." And that single loop which connects the two GBs provides an essential change of sign to the potential, so that nucleon binding can occur.

12.2 Formulation

We now generalize the initial steps of the previous $Q - \bar{Q}$ scattering to the situation of a triad of quarks bound into a singlet nucleon scattering with another such nucleon. For each nucleon there will occur the product of three $G_c(x, y|gA)$ such as those written in (9.20) each with the same fraction of that nucleon's momentum, and with their color weightings Ω^a restricted so as to insure that the three bound quarks comprising each nucleon remain in a color-singlet state.

We replace the description of that combination by that of a nucleon, of momentum p and effective color weighting $\bar{\Omega}^a$, so defined such that only those combinations of Gell-Mann matrices of each of the basic quarks corresponds to gluons which may be exchanged and so preserve each triad of quarks in its bound, color-singlet state. Note that all such color-singlet exchanges can be absorbed and emitted by the quark line comprising the loop $\mathbf{L}[A]$, for the Fradkin representation of $\mathbf{L}[A]$ contains a trace over all possible combinations of color coordinates, By this simplification, we replace the essence of a 6-body problem by a 2-body problem; and this has the consequence that our subsequent estimation of the nucleon-nucleon binding potential produces a qualitative description of how nuclear forces can arise from basic QCD.

We next refer the reader to Section 2 of Chapter 9 and in particular to the functional operations of its with attention drawn to the translation operator $\exp[-\int \mathcal{Q}\cdot(gf\cdot\chi)^{-1}\cdot\frac{\delta}{\delta\Lambda}]$ acting upon $\exp\{\mathbf{L}[A]\}$ in (9.35). Here, $\mathcal{Q}$ refers to the coordinates of the two nucleons, and the translation operator inserts that dependence in a well-defined way into the $\mathbf{L}[A]$ written in Equation (9.35) of that Chapter. The remaining linkage operator of Equation (9.35) generates a functional cluster expansion, discussed and derived in [Fried (1990)]; for our purposes, involving but a single loop and suppressing any gluon bundles exchanged across that loop, this remaining linkage operation can be neglected. This simplified analysis, which will require a simple renormalization, is sufficient to produce a reasonable, qualitative nuclear potential from basic 'realistic' QCD. By 'realistic', is meant a formulation of QCD which contains from its inception the idea that asymptotic quarks and/or antiquarks are found only in bound states, and hence their transverse coordinates cannot be specified exactly. In this Chapter, such transverse imprecision follows from the defining arguments found in Chapter 10.

We now ask the reader to imagine that functional derivatives are taken with respect to six pairs of $\eta, \bar{\eta}$ sources, "bringing down" six $\mathbf{G}_c(x^{(i)}, y^{(i)}|A)$, which we divide into two groups of three,

$$\mathbf{G}_c(x_I^{(1)}, y_I^{(1)}|A)\mathbf{G}_c(x_I^{(2)}, y_I^{(2)}|A)\mathbf{G}_c(x_I^{(3)}, y_I^{(3)}|A) \tag{12.1}$$

for nucleon I, and a similar triad with coordinate superscripts (4), (5), and (6) for nucleon II, beginning the computation as if we were calculating a six-quark scattering amplitude. Each $\mathbf{G}_c[A]$ will bring to its triad the A-dependence associated with an eikonal/high-energy limit of its exact Fradkin representation, of form

$$\exp\left[-ig\frac{p^{(i)}_{\mu(I,II)}}{m}\int_{-\infty}^{+\infty} ds\Omega^{a,(i)}_{(I,II)}(s)A^a_\mu(y^{(i)}_{(I,II)} - s\frac{p^{(i)}_{(I,II)}}{m})\right]. \tag{12.2}$$

We now introduce the bound-state nature of each triad of quarks by first suppressing the coordinate superscripts, $y^{(i)}_{(I,II)} \to y_{(I,II)}$, and $p^{(i)}_{(I,II)} \to (1/3)p_{(I,II)}$, since each quark must have essentially the same space-time and momentum coordinates if its nucleon is to remain intact. This means that the product of the three factors of (12.2) which are now a property of each nucleon may be written as

$$\exp\left[-i\frac{g}{3}\frac{p_{\mu,(I,II)}}{m}\int_{-\infty}^{+\infty}ds\sum_{i=1}^{3}\Omega^{a,(i)}_{(I,II)}(s)A^a_\mu\Big(y_{(I,II)}-\frac{s}{3}\frac{p_{(I,II)}}{m}\Big)\right]. \quad (12.3)$$

We emphasize that (12.3) refers to the A-dependence of each nucleon after the linkage operations binding each triad of quarks have been performed, which binding analysis we here suppress. Further, for each nucleon to remain bound for all (proper) times, there must exist a relation between the $\Omega^{a,(i)}(s)$ such that only color singlets are exchanged between nucleons I and II, and this relation should be independent of s. Since the $\Omega^{a,(i)}$ define the Gell-Mann matrices $\lambda^{a,(i)}$ inserted between initial and final nucleon states, there must be a relation between the $\Omega^{a,(i)}$ guaranteeing that each nucleon remains a color singlet. We thus simplify (12.3) by introducing $\bar{\Omega}^a_{(I,II)} = \Sigma^3_{i=1}\Omega^{a,(i)}_{(I,II)}$, and re-scaling $s \to 3s$, so that (12.3) becomes

$$\exp\left[-ig\frac{p_{\mu,(I,II)}}{m}\bar{\Omega}^a_{(I,II)}\int_{-\infty}^{+\infty}ds\ A^a_\mu\Big(y_{(I,II)}-s\frac{p_{(I,II)}}{m}\Big)\right]. \quad (12.4)$$

The modification, representing the "realistic" QCD defined and used in the two preceding Chapters, replaces in the exact Fradkin representation each $A^a_\mu(y'-u(s'))$ by

$$\int d^2y'_\perp a(y_\perp - y'_\perp)A^a_\mu(y'-u(s')), \quad (12.5)$$

where $y'_\mu = (iy_0; \vec{y}'_\perp, y_L)$ and $y_\mu = (iy_0; \vec{y}_\perp, y_L)$ represents the coordinate of an individual nucleon. Since we are assuming strict binding of each nucleon, the only transverse imprecision we must specify for this analysis is that between the quarks of one nucleon and those of the other; and since we have assumed that such gluon exchanges are not in any way to disrupt the binding of quarks within each nucleon, we shall invoke transverse imprecision for the y'_μ coordinates of $A^a_\mu(y'-u(s'))$ of (12.5), replacing (12.4) by

$$\exp\left[-i\int d^4w\ Q^a_{\mu,(I,II)}(w)A^a_\mu(w)\right] \quad (12.6)$$

where

$$\mathcal{Q}^a_{\mu,(I,II)}(w) = g\left(\frac{p_{\mu,(I,II)}}{m}\right)\bar{\Omega}^a_{(I,II)} \times \int d^2y'_\perp a(y_{\perp,(I,II)} - y'_{\perp,(I,II)})\delta^{(4)}\left(w - y'_{(I,II)} + s\frac{p_{(I,II)}}{m}\right). \tag{12.7}$$

In so doing, we have replaced the Fradkin coordinates $u_\mu(s)$ by the eikonal combinations $\frac{sp_I}{m}$ or $\frac{sp_{II}}{m}$, and have neglected quark spin dependence. The binding process effectively transforms each triad of quark Green's functions into a single nucleon Green's function of mass M and 4-momentum p_μ where the exponential (12.6), the effective 'relic' of its original three quarks, is retained for subsequent use in calculating the interaction between both nucleons. Each nucleon's Green's function now contributes to the eikonal scattering amplitude of the two nucleons, and in their CM takes on the standard form

$$\mathcal{T}(s,t) = \frac{is}{2M^2}\int d^2\, e^{i\vec{q}\cdot\vec{b}}[1 - e^{i\mathbb{X}(b,s)}], \tag{12.8}$$

where $\vec{q}$ is the momentum transfer of this scattering process, $s = -(p_I + p_{II})^2$, $t = -(p_I - p'_{II})^2 = -\vec{q}^{\,2}$, and where

$$e^{i\mathbb{X}(b,s)} = \mathcal{N}\int d[\chi]e^{\frac{i}{4}\int\chi^2} \cdot e^{\mathcal{D}_A}\cdot e^{\frac{i}{4}\int\chi\cdot\mathbf{F}+\frac{i}{2}\int A\cdot(\mathbf{D}_c^{-1})\cdot A}e^{-i\int(\mathcal{Q}_I+\mathcal{Q}_{II})\cdot A}e^{\mathbf{L}[A]}\Big|_{A\to 0} \tag{12.9}$$

with normalization constant $\mathcal{N}$ defined such that $\mathbb{X}\to 0$ for $g\to 0$.

The linkage operation (12.9) then has the Gaussian form

$$e^{-\frac{i}{2}\int\frac{\delta}{\delta A}\cdot D_c\cdot\frac{\delta}{\delta A}}\cdot e^{\frac{i}{2}\int A\cdot K\cdot A+i\int\mathcal{R}\cdot A}e^{\mathbf{L}[A]}\Big|_{A\to 0}, \tag{12.10}$$

where $K^{ab}_{\mu\nu} = gf^{abc}\chi^c_{\mu\nu} + (\mathbf{D}_c^{-1})^{ab}_{\mu\nu}$ and $\mathcal{R}^a_\mu = \partial_\nu\chi^a_{\mu\nu} - \mathcal{Q}^a_{I,\mu} - \mathcal{Q}^a_{II,\mu}$. As in the previous Chapters, the functional operation may be carried through exactly, yielding for (12.10)

$$e^{-\frac{i}{2}\int\mathcal{R}\cdot(gf\cdot\chi)^{-1}\cdot\mathcal{R}+\frac{1}{2}\mathbf{Tr}\ln(gf\cdot\chi)^{-1}} \cdot e^{+\frac{i}{2}\int\frac{\delta}{\delta A}\cdot(gf\cdot\chi)^{-1}\cdot\frac{\delta}{\delta A}}\cdot e^{-\int\mathcal{R}\cdot(gf\cdot\chi)^{-1}\cdot\frac{\delta}{\delta A}}\cdot e^{\mathbf{L}[A]}\Big|_{A\to 0}. \tag{12.11}$$

The first line of (12.11) may be rewritten as

$$[\det(gf\cdot\chi)]^{\frac{1}{2}}\cdot\exp\left\{-\frac{i}{2}\int\left[\partial\chi-\mathcal{Q}_I-\mathcal{Q}_{II}\right]\cdot(gf\cdot\chi)^{-1}\cdot\left[\partial\chi-\mathcal{Q}_I-\mathcal{Q}_{II}\right]\right\}, \tag{12.12}$$

and we here rely on the strong coupling limit of $g \gg 1$, keeping only the terms $\mathcal{Q}_{I,II}$ (this is really not necessary, but it simplifies the analysis; if the $\partial\chi$-terms are retained, the normalization integrals become more complicated, but the thrust of the procedure is the same).

Furthermore, the terms of (12.12) proportional to two factors of $\mathcal{Q}_I$ and to two factors of $\mathcal{Q}_{II}$ are "self-energy" corrections to the respective nucleon propagators, and they will be suppressed, since we are here interested only in the interaction of one nucleon upon the other. A similar remark may be made for those terms containing a single factor of $\partial\chi$ and either $\mathcal{Q}_I$ or $\mathcal{Q}_{II}$, for they correspond to "tadpole"-like structures attached to either nucleon, and are not relevant here. With these simplifications, (12.12) is replaced by

$$[\det(gf\cdot\chi)]^{\frac{1}{2}}\cdot\exp\left\{-i\int\mathcal{Q}_I\cdot(gf\cdot\chi)^{-1}\cdot\mathcal{Q}_{II}\right\}, \tag{12.13}$$

which, except for different color factors, has the form of the eikonal function describing the interaction between a pair of quarks.

For the impact parameter range between nucleons in which we are interested, it turns out that (12.13) gives an unimportant contribution to the nucleon-nucleon potential; for simplicity, we here neglect it, in contrast to the true source of that potential, which arises from the action of the linkage/diplacement operators of (12.11) upon $\exp\{\mathbf{L}[A]\}$.

Denoting the linkage operator of (12.11) by

$$\exp[\bar{\mathcal{D}}_A]=\exp\left[-\frac{i}{2}\int\frac{\delta}{\delta A}\cdot(-gf\cdot\chi)^{-1}\cdot\frac{\delta}{\delta A}\right], \tag{12.14}$$

where $(-gf\cdot\chi)^{-1}$ represents each non-perturbative gluon bundle to be exchanged between the quark lines which form the closed loop $\mathbf{L}[A]$, its action upon $\mathbf{L}[A]$ is most conveniently described in terms of a functional cluster-decomposition as

$$e^{\bar{\mathcal{D}}_A}\cdot e^{\mathbf{L}[A]}=\exp\left[\sum_{n=1}^{\infty}\frac{1}{n!}\bar{\mathbf{L}}_n\right],\quad \bar{\mathbf{L}}_n=e^{\bar{\mathcal{D}}_A}\cdot(\mathbf{L}[A])^n\big|_{\text{connected}}, \tag{12.15}$$

where "connected" requires at least one gluon bundle exchanged between different $\mathbf{L}[A]$'. In this Chapter we shall be concerned only with the simplest possible application of a single closed loop, and for this we may suppress the linkage operation of (12.15); while retaining the functional displacement operation of (12.11). With these simplifications, the second line of (12.11) becomes

$$\exp\{\mathbf{L}[A-(gf\cdot\chi)^{-1}\cdot(\mathcal{Q}_I+\mathcal{Q}_{II})]\}\big|_{A\to 0} = \exp\{\mathbf{L}[-(gf\cdot\chi)^{-1}\cdot(\mathcal{Q}_I+\mathcal{Q}_{II})]\}, \tag{12.16}$$

and our eikonal simplifies to

$$e^{i\mathbb{X})(b,s)}\mathcal{N}\int d[\chi]e^{\frac{i}{4}\int\chi^2}\cdot[\det(gf\cdot\chi)]^{-\frac{1}{2}}\cdot e^{\mathbf{L}[-(gf\cdot\chi)^{-1}\cdot(\mathcal{Q}_I+\mathcal{Q}_{II})]}. \tag{12.17}$$

In order to calculate the vacuum loop contribution to (12.17), we first write a Fradkin Representation for $\mathbf{L}[A]$,

$$\begin{aligned} \mathbf{L}[A] = -\tfrac{1}{2}\ \mathbf{Tr}\int_0^\infty \frac{dt}{t}e^{-itm^2}\mathcal{N}(t)\int d[v]e^{\frac{i}{2}\int_0^\ell dt' v\cdot(2h)^{-1}\cdot v}\delta^{(4)}(v(t)) \\ \times\int d^4x\int d[\hat{\alpha}]\int d[\hat{\Omega}]e^{i\int\ dt'\hat{\Omega}^a(t')\hat{\alpha}^a(t')}\left(e^{i\int_0^t dt'\hat{\alpha}^b(t')\lambda^b}\right)_+ \\ \times\left[e^{-ig\ \int_0^t\ dt'\ v'_\alpha(t')\hat{\Omega}^a(t')\ \int\ d^2x_\perp\ a(x_\perp x'_\perp)A^a_\alpha(x'-v(t'))} - 1\right], \end{aligned} \tag{12.18}$$

where $x'_\mu = (ix_0;\vec{x}'_\perp, x_L)$, $h(t_1,t_2) = \theta(t_1-t_2)t_2 + \theta(t_2-t_1)t_1 = \frac{1}{2}(t_1 + t_2 - |t_1-t_2|)$, $\mathcal{N}(t)$ is the normalization for the Gaussian functional integral over $v_\alpha(t')$, $\mathbf{Tr}$ denotes a trace over Dirac and color indices, and the hat notation of $\hat{\alpha}$ and $\hat{\Omega}$ is used to distinguish these loop color-variables from those of nucleons I and II. Again, in the interests of simplicity, we shall neglect all spin dependence of the quark loop, and, for clarity, have chosen the longitudinal and transverse directions of the loop to lie in the respective directions defined by the nucleons in their CM.

With the simplifications of the last two paragraphs, all of the structure that remains in our eikonal amplitude arises from that nucleon dependence, $\mathcal{Q}_I$ and $\mathcal{Q}_{II}$, which has been translated into the argument o $\mathbf{L}$ in (12.17), as its argument A is shifted by the amount $-(gf\cdot\chi)^{-1}\cdot(\mathcal{Q}_I+\mathcal{Q}_{II})$. But this shift occurs in the exponential factor of (12.18), whose expansion corresponds to multiple quark loops exchanged between the nucleons. The simplest, and probably the most important effect arises from the 'exchange of a single quark loop, proportional to the factor $\mathcal{Q}_I$ multiplying $\mathcal{Q}_{II}$, which contribution may be extracted from the quadratic expansion of that exponential factor, neglecting tadpole and self-energy corrections to the nucleons. We therefore replace the third line of (12.18) by

$$
\begin{aligned}
+\, g^2 & \int_{-\infty}^{+\infty} ds_1 (\frac{p_{I,\mu}}{m}) \int_{-\infty}^{+\infty} ds_2 \left(\frac{p_{II,\nu}}{m}\right) \bar{\Omega}^a \bar{\Omega}^b \int_0^t dt_1 v'_\alpha(t_1) \hat{\Omega}^c(t_1) \\
& \int_0^t dt_2 v'_\beta(t_2) \hat{\Omega}^d(t_2) \cdot \int d^2x'_\perp a(x_\perp - x'_\perp) \int d^2x''_\perp a(x_\perp - x''_\perp) \\
& \int d^2y'_{I,\perp} a(y_{I,\perp} - y'_{I,\perp}) \int d^2y'_{II,\perp} a(y_{II,\perp} - y'_{II,\perp}) \\
& \cdot\ \delta^{(4)}(y'_I - s_1 \frac{p_1}{m} - x' + v(t_1)) \cdot \delta^{(4)}(y'_{II} - s_2 \frac{p_{II}}{m} - x'' + v(t_2)) \\
& \cdot\ (f \cdot \chi(y'_I - s_1 \frac{p_I}{m})^{-1} \Big|_{\mu\alpha}^{ac} \cdot (f \cdot \chi(y'_{II} - s_2 \frac{p_{II}}{m})^{-1} \Big|_{\mu\alpha}^{db},
\end{aligned} \tag{12.19}
$$

where the generic notation of $z'_\mu = (iz_0; z'_\perp, z_L)$ for any x'_μ or y'_μ is understood. Here, x_μ represents the loop coordinate, which, along with the functional integration of the first line of (12.18) will be performed shortly. Notice that various factors of 2, $-i$, and g, have been combined to produce the coefficient $+g^2$ multiplying (12.19), and that this translated approximation to $\mathbf{L}[A]$ has become the essence of the desired eikonal function, at least before the needed Halpern integrations are performed.

It will be most convenient to choose the zero of the coordinates $y_{I,0}$ and $y_{II,0}$ at the instant of their CM distance of closest approach, which then corresponds to $s_1 = s_2 = 0$. The argument of each inverse $(f \cdot \chi)$ is then independent of proper time; consider, e.g., the combination $(y'_I - s_1 \frac{p_I}{m})_\mu = (y_{I,0} - s_1 \frac{E}{m}; y'_{I,\perp} - s_1 \frac{p_\perp}{m}, y_{I,L} - s_1 \frac{p_L}{m})$. But in the CM, $y_{I,0} = s_1 \frac{E}{m}, p_\perp \simeq 0$, and for large momenta $y_L \sim y_0$ as $p_L \sim E$, so that $(y'_I - s_1 \frac{p_I}{m})_\mu$ reduces to $(0; y'_{I,L}, 0)$. The same argument, with the CM signs of $p_{I,L}$ and $y_{I,L}$ reversed, holds for the combination $(y'_{II} - s_2 \frac{p_{II}}{m})_\mu \to (0; y'_{II,\perp}, 0)$.

This now allows the $s_{1,2}$ integrals to be performed, and for this a Fourier representation of the two delta functions of (12.19) is convenient, which yields

$$
\begin{aligned}
m^2 & \int \frac{d^4k}{(2\pi)^4} \int \frac{d^4q}{(2\pi)^4} (2\pi)^2 \delta(q \cdot p_I) \delta(k \cdot p_{II}) \\
& \cdot\ e^{iq\cdot(y'_I - x') + ik\cdot(y'_{II} - x'')} \cdot e^{iq\cdot v(t_1) + ik\cdot v(t_2)},
\end{aligned} \tag{12.20}
$$

and where $\delta(q \cdot p_I) = \frac{1}{E}\delta(q_0 - q_L \frac{p_L}{E}) \simeq \frac{1}{E}\delta(q_0 - q_L)$, $\delta(k \cdot p_{II}) = \frac{1}{E}\delta(k_0 + k_L \frac{p_L}{E}) \simeq \frac{1}{E}\delta(k_0 + k_L)$, $p_L = p_{IL} = -p_{IIL}$, and $E = E_1 =$

E_{II}. Then, the integrations $\int dx_0 \int dx_L$ produce the additional factors $(2\pi)^2\delta(q_0+k_0)\delta(q_L+k_L)$, which multiply the previous line, and produce a net combination of $\frac{1}{2}\frac{(2\pi)^2}{E^2}\delta(q_0)\delta(k_0)\delta(q_L)\delta(k_L)$, so that the remaining q- and k-integrals refer to transverse components only.

Before performing the transverse integrations, it will be convenient to make one further simplification, one which appears as a reasonable approximation, but can be justified following the argument of Appendix E. This simplification replaces the arguments of each inverse $(f \cdot \chi)$ factor by their "expected" values $y_{I,\perp}$ and $y_{II\perp}$. This step would appear to be a reasonable approximation because the $a(y_{I\perp}-y'_{I\perp})$ and $a(y_{II\perp}-y'_{II\perp})$ distributions are each peaked about a zero value of their arguments, which essentially forces the primed transverse $y'_\perp$ coordinates to lie close to their unprimed values. But if appropriate care is taken in the evaluation of the χ-integrations, one eventually finds the same form of result as when this simplification is first performed. Hence, in the interest of clarity and simplicity, we now adopt the replacements: $\chi(y'_\perp) \to \chi(y_\perp)$.

We next evaluate the multiple transverse integrals of (12.19) by first writing Fourier transforms for each of the $a(z_\perp))$ distributions,

$$a(z_\perp) = \int \frac{d^2k}{(2\pi)^2}\tilde{a}(k)e^{ik\cdot z_\perp}, \tag{12.21}$$

so that

$$\begin{aligned}
(2\pi)^{-2} & \int d^2q \int d^2k \int d^2x'_\perp \int d^2x'_\perp \int d^2x''_\perp \int d^2y'_{I,\perp} \int d^2y'_{II,\perp} \\
& \cdot\; a(x_\perp - x'_\perp)a(x_\perp - x''_\perp)a(y_{I\perp} - y'_{I\perp})a(y_{II\perp} - y'_{II\perp}) \\
& \cdot\; e^{iq\cdot(y'_{I\perp}-x'_\perp)+ik\cdot(y'_{II\perp}-x''_\perp)} \\
&= (2\pi)^{-2} \int d^2q|\tilde{a}(q)|^2|\tilde{a}(q)|^2 e^{iq\cdot B} \\
&= \int d^2b\varphi(b)\varphi(\vec{B}-\vec{b}) \equiv \bar{\varphi}(B),
\end{aligned} \tag{12.22}$$

where $\bar{\varphi}(B)$ will provide a slower fall-off with increasing B than does $\varphi(b)$ with increasing b. Here, $B = y_{I\perp} - y_{II\perp}$, and $\varphi(b)$ is the modified statement of transverse imprecision introduced in [Fried, et al. (2010)] and made precise in (11.12): $\varphi(b) \simeq \frac{\mu^2}{\pi}e^{-(\mu b)^{2+\xi}}$, $\xi \ll 1$.

The integral of (12.22) is not the final statement of B dependence, because a term proportional to $q_\alpha q_\beta$ arising from the evaluation of the functional integral of (12.18) and appearing in (12.26) must still be included.

One requires

$$\mathcal{N} \int d[v] e^{\frac{i}{2}\int v\cdot(2h)^{-1}\cdot v}\; v'_\alpha(t_1) v'_\beta(t_2) \delta^{(4)}(v(t)) e^{iq\cdot(v(t_1)-v(t_2))}, \tag{12.23}$$

which may be accomplished by inserting a Fourier representation of $\delta^{(4)}(v(t))$, and rewriting (12.23) as

$$\frac{1}{i}\frac{\partial}{\partial t_a}\frac{\delta}{\delta g_\alpha(t_a)}\frac{1}{i}\frac{\partial}{\partial t_b}\frac{\delta}{\delta g_\beta(t_b)} \cdot \mathcal{N} \int d[v] e^{\frac{i}{2}\int v\cdot(2h)^{-1}\cdot v + i\int_0^t dt'\; v_\mu(t')[f_\mu(t')+g_\mu(t')]}\Big|_{t_a\to t_1, t_b\to t_2} \tag{12.24}$$

where $f_\mu(t') = p_\mu \delta(t'-t)$, and $g_\mu(t') = q_\mu[\delta(t'-t_1) - \delta(t'-t_2)]$. The normalized, Gaussian functional integral of (12.24) is then

$$\exp\{-i\int_0^t dt' \int_0^t dt''[f_\mu(t)+g_\mu(t')]\; h(t',t'')[f_\mu(t'')+g_\mu(t'')]\}, \tag{12.25}$$

and the functional and conventional derivatives of (12.24), as well as the resulting Gaussian $\int d^4p$ are immediate. Combining all factors, one obtains for the translated and simplified $\mathbf{L}[A]$ (12.18) the result

$$\frac{i}{4}\left(\frac{p_{I\mu}p_{II\nu}}{E^2}\right)\frac{g^2}{(4\pi)^2}\int d^2q|\tilde{a}(q)|^2|\tilde{a}(q)|^2\; e^{iq\cdot B}\int_0^\infty \frac{dt}{t}e^{-itm^2}\bar{\Omega}^a\bar{\Omega}^b$$
$$\cdot \mathcal{N}'\int d[\hat{\alpha}]\int d[\hat{\Omega} e^{i\int_0^t dt'\hat{\alpha}^a(t')\hat{\Omega}^a(t')}\mathbf{Tr}(e^{i\int_0^t dt'\hat{\alpha}^a(t')\lambda^a})_+$$
$$\cdot \int_0^1 d\mathcal{Z}_1 \int_0^1 d\mathcal{Z}_2\left\{\frac{2i}{t}\delta_{\alpha\beta}[\delta(\mathcal{Z}_1-\mathcal{Z}_2)-1] + q_\alpha q_\beta[1-4(\mathcal{Z}_1-\mathcal{Z}_2)^2]\right\}$$
$$\hat{\Omega}^c(\mathcal{Z}_1 t)\hat{\Omega}^d(\mathcal{Z}_2 t)\cdot(f\cdot\chi(y_{I\perp}))^{-1}\big|^{ac}_{\mu\alpha}\cdot(f\cdot\chi(y_{II\perp}))^{-1}\big|^{db}_{\beta\nu}, \tag{12.26}$$

where we have replaced $t_{1,2}$ by $t\cdot\mathcal{Z}_{1,2}$, and $\int_0^t dt_{1,2}$ by $t\int_0^1 d\mathcal{Z}_{1,2}$.

Equation (12.26) is noteworthy for several reasons, among which is the special way in which the manifest gauge invariance of $\mathbf{L}[A]$ is displayed in the automatic cancelation of the quadratic divergence associated with the removal of the $\int_0 \frac{dt}{t^2}$ of (12.26). In Feynman graph language this does not happen automatically, for the divergence of the fermion loop "overpowers" the gauge invariance of the basic theory; and one must resort to other measures to remove that quadratic divergence. As Schwinger pointed out long ago [Schwinger (1951)], in his functional development of radiative corrections to QED in terms of proper time variables, such unwanted and improper terms never appear in calculations so defined.

The gauge-invariant divergence of this loop is logarithmic, as expected; and its renormalization displays the behavior associated with the property of "anti-shielding", as expected in QCD, rather than the "shielding" of QED. This divergence, associated with the lower limit of 0 in the t-integral of (12.26), may be described in configuration space by replacing that lower limit by a small quantity ϵ, of dimensions of (length)^2; in momentum space, this would correspond to a cut-off of $\Lambda^2 = 1/\epsilon$. It will be convenient to perform the variable change $t = \epsilon r$, and then rotate contours $r \to -iz$, so that the t-integral of (12.26) becomes

$$\int_1^\infty \frac{dz}{z} e^{-\frac{z}{\Lambda^2}[m^2+q^2|\mathcal{Z}_{12}|(1-|\mathcal{Z}_{12}|)]} \hat{\Omega}^c\left(\frac{-iz}{\Lambda^2}\mathcal{Z}_1\right)\hat{\Omega}^d\left(\frac{-iz}{\Lambda^2}\mathcal{Z}_2\right)$$
$$\simeq \left\{ \ln\left(\frac{\Lambda^2}{m^2}\right) - \ln\left[1 + \frac{q^2}{m^2}|\mathcal{Z}_{12}|(1-|\mathcal{Z}_{12}|)\right]\right\}\hat{\Omega}^c(0)\hat{\Omega}^d(0), \quad (12.27)$$

where $\mathcal{Z}_{12} = \mathcal{Z}_1 - \mathcal{Z}_2$, and we have allowed Λ to become arbitrarily large in the arguments of $\hat{\Omega}^c$ and $\hat{\Omega}^d$; we have also, for the moment, suppressed the t-dependent integrals coupling $\hat{\alpha}^a$ and $\hat{\Omega}^d$ to the Gell-Mann matrices λ^c. The renormalized coupling of this order g^2 bundle diagram may be defined by the relation suggested by (12.27), as

$$g_R^2(q^2) = g^2 \ln\left[\frac{\Lambda^2}{m^2 + q^2|\mathcal{Z}_{12}|(1-|\mathcal{Z}_{12}|)}\right] \quad (12.28)$$

which displays the expected QCD form, of an effective, or (partially) renormalized coupling that decreases with increasing momentum transfer. And since the q-values expected from its subsequent integration are less than the quark mass, and both are understood to be far less than any realistic cut-off adopted for Λ, (12.28) may be most simply approximated by

$$g_R^2 = g^2 \ln\left(\frac{\Lambda^2}{m^2}\right), \quad (12.29)$$

where it is clear that the bare coupling g of the original Lagrangian is smaller than the renormalized coupling, in contrast to Abelian QED, where the reverse holds.

From this example one sees that our formalism is non-perturbative in the sense of summing over all gluon exchanges between specified quarks; but that if one of those quark lines is part of a closed loop, then a perturbation expansion can be defined involving increasing numbers of gluon bundles exchanged between that closed loop and other, specified quarks, which may themselves be associated with other quark loops. Can the non-perturbative nature of our analysis be extended to include all possible $\mathbf{L}[A]$ interactions? We hope to answer this very non-trivial question in a subsequent publication.

The color dependence of (12.19) remains to be treated, and for this it is simplest to return to that stage of calculation before renormalization was discussed. There, the factors of $\hat{\Omega}^c(t_1)$ and $\hat{\Omega}^d(t_2)$ remain to be evaluated, which process consists of converting them into Gell-Mann matrices λ^c and λ^d. It can easily be shown that the commuting factors of $\hat{\Omega}^c(t_1)\hat{\Omega}^d(t_2)$ are to be replaced by $\lambda^c\lambda^d\theta(t_1-t_2)+\lambda^d\lambda^c\theta(t_2-t_1)$, while, simultaneously, the functional integrations over $\hat{\alpha}$ and $\hat{\Omega}$ have disappeared.

After renormalization, in which the $\int dt$ is effectively evaluated close to its lower limit, and where $t_{1,2} \Rightarrow t\cdot \mathcal{Z}_{1,2}$, as $t \to 0$, $\theta(t_1-t_2) \to \theta(t_2-t_1) \to \theta(0) = 1/2$ and the product $\hat{\Omega}^c(t_1)\hat{\Omega}^d(t_2)$ is replaced by $\frac{1}{2}\{\lambda^c, \lambda^d\}$. As noted above, for simplicity and ease of presentation, we have neglected quark spin dependence, and its associated λ-dependence, so that the $\mathbf{Tr}$ operation over both Dirac and color indices yields $\mathbf{Tr}[\frac{1}{2}\{\lambda^c, \lambda^d\}] = 8\delta^{cd}$. Equation (12.26) then reduces to

$$\left(\frac{p_{I\mu}p_{II\nu}}{E^2}\right)\frac{g_R^2}{3\pi^2}\bar{\Omega}^a\bar{\Omega}^b\,(f\cdot\chi(y_{I\perp}))^{-1}\Big|^{ac}_{\mu\alpha}\cdot(f\cdot\chi(y_{II\perp}))^{-1}\Big|^{cb}_{\beta\nu}$$
$$\cdot\left(-\frac{\partial}{\partial B_\alpha}\frac{\partial}{\partial B_\beta}\right)\bar{\varphi}(B), \tag{12.30}$$

where the $q_\alpha q_\beta$ factors of (12.26) have been replaced by $\left(-\frac{\partial}{\partial B_\alpha}\frac{\partial}{\partial B_\beta}\right)$.

The relevant space-time indices enter here in the form

$$\left(\frac{p_{I\mu}p_{II\nu}}{E^2}\right)(f\cdot\chi(I))^{-1}\Big|^{ac}_{\mu\alpha}\cdot(f\cdot\chi(II))^{-1}\Big|^{cb}_{\beta\nu}, \tag{12.31}$$

and, remembering the antisymmetry of each element's color and space-time indices, and that the α, β are transverse indices, (12.31) may be rewritten as

$$+i(f\cdot\chi(I))^{-1}\Big|^{ac}_{4\alpha}\cdot(f\cdot\chi(II))^{-1}\Big|^{bc}_{\beta L}-i(f\cdot\chi(I))^{-1}\Big|^{ac}_{L\alpha}\cdot(f\cdot\chi(II))^{-1}\Big|^{bc}_{4\beta}, \tag{12.32}$$

where, because the longitudinal and energy components are far larger than the transverse momenta, the μ, ν indices correspond to 0 and L only. Using the Minkowski metric, where $\chi^a_{4\alpha} \equiv i\chi^a_{0\alpha}$, then $(f\cdot\chi)^{-1}\big|_{4\alpha} = -i(f\cdot\chi)^{-1}\big|_{0\alpha}$.

Further, in the CM system, where the longitudinal projections of y_I and y_{II} point in exactly opposite directions, while the χ variables depend only upon their respective transverse coordinates; then, the CM longitudinal projections of such χ^{-1} will point in opposite directions. In order to have similar, if arbitrary, constructions of the $\chi(I)$ and $\chi(II)$, we set $\chi'_{\beta L}(II) = -\chi(II)\big|_{\beta L}$, so as to bring (12.32) into the form

$$-\left\{(f \cdot \chi(I))^{-1}\big|_{\alpha 0}^{ca} \cdot (f \cdot \chi'(II))^{-1}\big|_{\beta L}^{cb} + (f \cdot \chi(I))^{-1}\big|_{\alpha L}^{ca} \cdot (f \cdot \chi(II))^{-1}\big|_{\beta 0}^{cb}\right\}. \tag{12.33}$$

The normalized integrals over $\int d^n\chi_{\beta L}(II)$ and $\int d^n\chi'_{\beta L}(II)$, are the same, and are unchanged; and since the values of $y_{I\perp}$ and$y_{II\perp}$ appearing in the arguments of each χ serve only to indicate that two separate integrations are required, one can interchange those arguments in the second term of (12.33) to obtain, in place of (12.33),

$$-\left\{(f \cdot \chi(I))^{-1}\big|_{\alpha 0}^{ca} \cdot (f \cdot \chi(II))^{-1}\big|_{\beta L}^{cb} + (f \cdot \chi(I))^{-1}\big|_{\beta 0}^{cb} \cdot (f \cdot \chi(II))^{-1}\big|_{\alpha L}^{ca}\right\}, \tag{12.34}$$

a result which is explicitly symmetric in a and b and in α and β.

As in a previous Chapter, we assume that each χ^c can be represented by an angular projection $\ddagger^c$ multiplying a magnitude R, $\chi^c = \mathcal{Z}^c R$, and we now suppress the result of those normalized angular integrals, assuming that the most significant behavior of our results is due to integration over the magnitudes. Of course, such a simplification must be checked by detailed, numerical calculation; but this would appear to be a reasonable approximation. Note that the index symmetries of (12.33) would be enforced by multiplication by $q_\alpha q_\beta$ of (12.26), and by the $\bar{\Omega}^a\bar{\Omega}^b$, corresponding to color singlet gluon emission and absorption of the two nucleons. There is then no difference between the two terms of (12.34); they are both going to give the same contribution, and so (12.34), after multiplication by the $\bar{\Omega}^{a,b}$, and $q_\alpha q_\beta$ is equivalent to

$$-2\bar{\Omega}^a\bar{\Omega}^b\, q_\alpha q_\beta\, (f \cdot \chi(I))^{-1}\big|_{0\alpha}^{ac} \cdot (f \cdot \chi(II))^{-1}\big|_{0\beta}^{bc}. \tag{12.35}$$

The attentive reader will notice that there is one aspect of our procedure of obtaining an effective potential from an eikonal function which remains to be discussed: what is to be done when the eikonal itself contains transverse components of coordinates or corresponding momentum transfer? Physically, each component of the initial momentum transfer q_μ of nucleon I must be transferred to the corresponding component of momentum transfer of nucleon II on the other side of the loop, $q_1(I)$ to $q_1(II)$ and $q_2(I)$ to $q_2(II)$; in other words, a $\delta_{\alpha\beta}$ must appear in (12.35), either from integrations over the "angular" components of the Halpern variables, which we have suppressed, or from the neglected quark spins, or as a definite statement of our procedure, which we now state: All such "free" indices are to be averaged over, a stipulation which has consequences in other contexts (renormalization theory and nuclear binding). In the present case, it means that $q_\alpha q_\beta$ is to be replaced by $\frac{1}{2}q^2\delta_{\alpha\beta}$ as is physically necessary. Then, we may write the simplified, normalized integrals to be performed as

$$\mathcal{N} \int_0^\infty dR_I\ R_I^3 \int_0^\infty dR_{II} R_{II}^3 e^{\frac{i}{4}(R_I^2+R_{II}^2)-i\frac{C(B,E)}{R_I R_{II}}}, \tag{12.36}$$

where $C(B,E) = \frac{1}{3}(\frac{g_R^2}{4\pi})(\delta^2)^2[(-\nabla_B^2)\cdot\bar{\varphi}(B)]$, and, as explained in detail in Section 9.3, δ^2 contains the scale change needed when passing from the Halpern FI to the individual $\int d^8\chi$: $\delta^2 = (\mu E)^{-1}$. The $R_{I,II}^3$, rather than the $R_{I,II}^7$, result from a factor $[R_{I,II}^8]^{-1/2}$ extracted from each determinantal factor $[\det(gf\cdot\chi)]^{\frac{1}{2}}$ of (12.13).

It is the double derivatives with respect to B, the impact parameter between the two nucleons, arising from the $q^2\delta_{\alpha\beta}$ components of the closed-loop integral, which provides the sign of a potential that produces nucleon binding; and it is in this qualitative possibility of generating a "model deuteron" from two bound nucleons that the possibility of obtaining true Nuclear Physics from transversally-averaged QCD appears.

12.3 A Qualitative Binding Potential

Before passing to the final steps of the calculation of this potential, we remind the reader that this treatment is based on the simplest possible realization of realistic QCD, based on a single, massive quark interacting with its complement of SU(3) massless gluons; flavors and electroweak interactions, as well as quark and nucleon spins and angular momenta have been neglected, for simplicity, and can be added separately, producing definite variations of the potential below. The rigorous property of Effective Locality, defined and discussed in detail in the previous three chapters immensely simplifies the original Halpern FI by reducing it, in the present case, to two sets of ordinary integrals; and we have here suppressed the "angular" color integrations, retaining dependence only on the magnitudes of the reduced Halpern variables, an approximation which must be verified by numerical calculation. Nevertheless, it should be of more than passing interest to see just how a qualitatively reasonable nucleon potential can appear from such basic QCD.

Of course, that potential is not meant to suggest that two neutrons will bind, for their fermionic nature has been suppressed with the neglect of their spins; nor would it be suggestive of two protons binding to form a nucleus, because both spin structure and electrodynamics have been omitted. That potential is not yet meant to be compared with precise experimental data, except in the sense of its qualitative behavior, producing for two distinguishable nucleons scattering at high relative energies, as well as the possibility of binding into a "model deuteron" at lower incident energies.

With the simplifications and approximations discussed in the preceding sections, we now write (12.9) in the form

$$e^{i\mathbb{X}(B,E)} = \mathcal{N} \int_0^\infty dR_I \, R_I^3 \int_0^\infty dR_{II} \, R_{II}^3 e^{\frac{i}{4}(R_I^2 + R_{II}^3) - i\frac{C(B,E)}{R_I \cdot R_{II}}}, \quad (12.37)$$

where we make the further, simplifying approximation of suppressing the parameter $\xi \simeq 0.1$, of $\varphi(b)$, which was crucial in the construction of quark binding, but would here only slightly change the shape of the nucleon binding potential. Setting then $\varphi(b) = (\mu^2/\pi)\exp[-(\mu b)^2]$, one finds

$$C(B,E) = \frac{g_R^2}{6\pi^2}\left(\frac{\mu}{E}\right)^2 [2 - \mu^2 B^2] e^{-\frac{\mu^2 B^2}{2}}. \quad (12.38)$$

There are several methods of obvious approximation to the integral of (12.37):

(1) A change of variables to polar coordinates, $R_I = R\sin(\theta)$, $R_{II} = R\cos(\theta)$, for which the radial integral can be done immediately, but the subsequent angular integral requires an approximation.
(2) Both $R_{I,II}$ integrands correspond to a function rising as $|R_{I,II}|$ increases from zero, and then falling away to 0 as these coordinates become large; and both may be approximated by (different) Gaussian approximations.
(3) Both (1) and (2) lead to rather complicated expressions involving fractional powers of complex functions. There is, however, a simpler approximation, available in this eikonal context where $C(B,E)$ contains

the factor $\delta^4\mu^4 = (\mu/E)^2$, where E is the CM energy of the scattering nucleons. In conventional eikonal representations, there is always a dimensionless, energy-dependent, kinematical factor, $\gamma(E)$, multiplying an impact-parameter-dependent function which is the "true" eikonal function, derived from an initial potential function; and if that combination is small, then the final eikonal amplitude may have its exponential factor expanded, so that only the linear dependence of that exponential is retained. Here, that energy-dependent $\delta^4\mu^4$-factor is surely small, but is it the correct $\gamma(E)$? In Potential Theory and in various forms of QFT, the functional form of $\gamma(E)$ can vary widely, but we have no precedent here to specify the "correct" form of $\gamma(E)$ to appear upon the exchange of a pair of gluon bundles supporting a quark loop. We shall therefore make the simplest choice of adopting $[\mu\delta(E)]^4$ as our tentative $\gamma(E)$; and at the very end of the calculation return to see if this choice is consistent with the order-of-magnitude of our qualitative potential.

We now expand to first order both the eikonal amplitude of (12.9), which is the left-hand-side of (12.37), and the exponential factor containing $C(B,E)$ of (12.37), so that

$$\begin{aligned} i\mathbb{X}(B,E) = i\mathbb{X}\cdot\gamma(E) &= -iC(B,E)\mathcal{N}J^2, \\ J &= \int_0^\infty dR\ R^3 e^{\frac{i}{4}R^2}, \\ \mathcal{N}\ J^2 &= -i\frac{\pi}{4} \end{aligned} \tag{12.39}$$

where the "true" eikonal function is

$$\mathbb{X} \simeq \left(\frac{i}{6}\right)\left(\frac{g_R^2}{4\pi}\right)[2-\mu^2B^2]\ e^{-\frac{\mu^2B^2}{2}}. \tag{12.40}$$

This situation differs from that of the quark binding calculation, where the large impact parameter of interest generated a large eikonal function, but a small amplitude; here, both the eikonal and the amplitude are small.

The relation between the eikonal function and the effective potential is

$$\mathbb{X} = -\int_{-\infty}^{+\infty} dz_L V(\vec{B} + \hat{\mathbf{P}}_L z_L) \tag{12.41}$$

and as explained in the previous Chapter, the eikonal is real for a purely scattering potential, $V = V_S$; but for a potential which can lead to binding, or to the production of other particles, the potential chosen must have the form $V = V_S - iV_B$, so that the eikonal which corresponds to binding is imaginary. The reason is unitarity, since if extra particles, or a new bound state can be produced, the amplitude of the initial state must be reduced. In our reversed situation, starting from the construction of a QCD amplitude, we find a clear signal of a binding potential, with V_B appearing as a real quantity,

$$\int_{-\infty}^{+\infty} dz_L V_B = \frac{1}{12}\frac{g_R^2}{4\pi}(2 - \mu^2 B^2)\ e^{-\mu^2 B^2/2}. \tag{12.42}$$

To obtain the effective potential one first calculates the two-dimensional Fourier transform of $-i\mathbb{X}(B)$, which can be expressed as proportional to

$$\left(\frac{\mu^2}{4\pi^2}\right)k_\perp^2 \int d^2 B e^{ik_\perp \cdot B - \mu^2 B^2/2}, \tag{12.43}$$

then continue $k_\perp^2$ to three dimensions, $k_\perp^2 \to k_\perp^2 + k_L^2$, and calculate the three-dimensional Fourier transform of (12.43); which yields, after removing the factor $\gamma(E) = (\mu\delta)^4$,

$$V(r) \simeq c\left(\frac{g_R^2}{4\pi}\right)\ \mu[2 - \mu^2 r^2]\ e^{-\frac{\mu^2 r^2}{2}}, \tag{12.44}$$

with $c = \frac{1}{6}(2\pi)^{-3/2}$. At high energies and small momentum transfers, this potential when multiplied by $(-i)$ corresponds to an effective scattering potential.

The form of this potential is sketched in Figure 12.3, and it will look familiar to those who have inferred a nucleon potential from experimental scattering data, starting with the potentials of the 1951 paper of [Jastrow (1951)][1]. It must be noted that this potential is not meant to be relevant at distances $\mu r < 1$, which is where the multiple gluon exchanges of the gluon bundles of Figure 12.2, as well as those of the omitted gluon-binding interactions of each triad of bound nucleons take place. And of course, we have neglected electromagnetic effects, as well as all spin and angular momentum modifications, which can be included in more detailed estimations.

We have two parameters at our disposal, the mass scale $\mu \simeq m_\pi$, and $g_R^2/4\pi$, which can be chosen so as to produce a ground state with a binding

[1]A very nice fit to the shape of the potential of Figure 12.3 of the present Chapter is the average of the singlet and triplet potentials of Figure 1 of this reference.

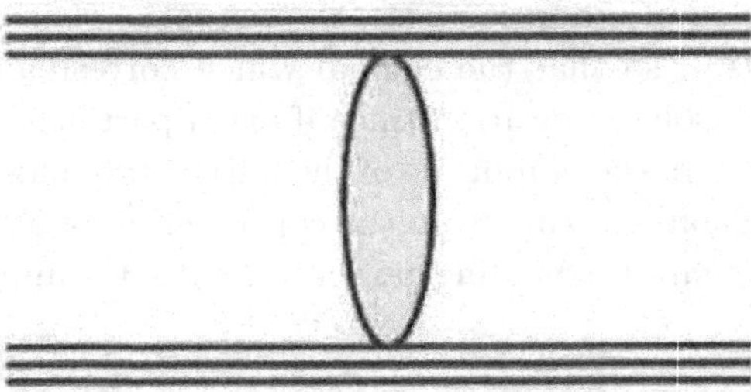

Fig. 12.1 A gluon bundle exchanged between two nucleons.

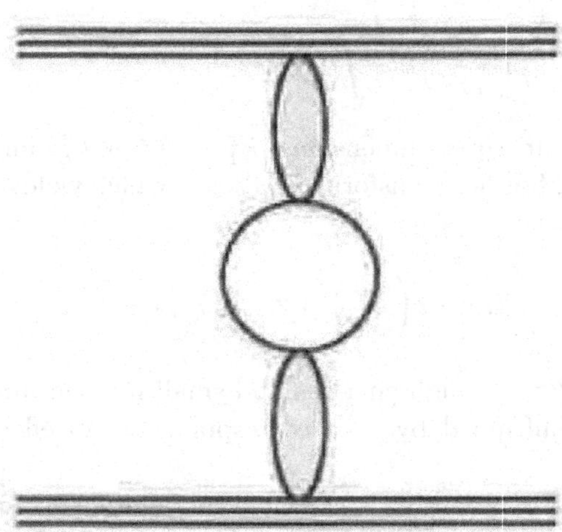

Fig. 12.2 Quark loop exchange through gluon bundles between two nucleons.

energy of -2.2 MeV. Of course, from the crudeness of the approximations made in our various estimations, we would be happy to obtain a binding energy to within a factor of 10 of this numerical result, but as it happens, we shall do somewhat better. The corresponding calculation is demonstrated in the next section, using the elementary Quantic technique [Balibar (1989)] of estimating a ground state. But, simplifying approximations aside, this is clearly a potential which can bind a pair of distinct, uncharged nucleons; and it is obtained analytically, from basic, transversally-averaged QCD.

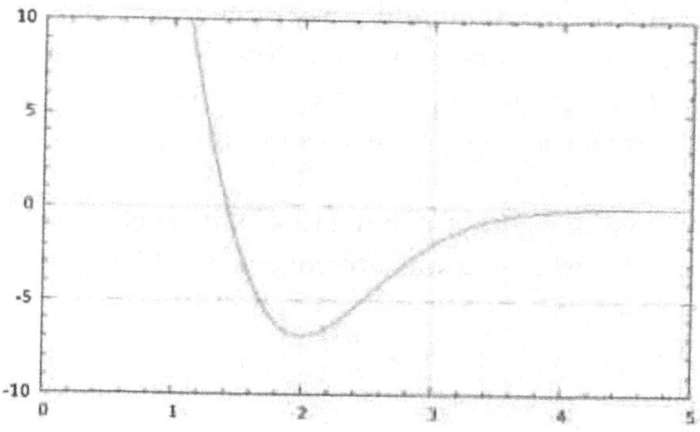

Fig. 12.3 Nucleon potential.

12.4 Binding Estimations

The non-relativistic Hamiltonian of two, equal mass particles interacting with the above potential is

$$E = \frac{p^2}{m} + V(r) \to \frac{1}{mr^2} + V(r). \tag{12.45}$$

One can write this non-relativistic energy in dimensionless form as

$$\frac{E(y)}{M} = \frac{1}{y^2} + \frac{V_0}{M}[2 - y^2]e^{-\frac{y^2}{2}}, \quad y = \mu r, \tag{12.46}$$

where we have set $\mu = m_\pi$. In units of MeV, $M = \mu^2/m = 18.2$, $g_R^2/4\pi = \aleph \cdot 10$, and, combining all the relevant factors of the previous paragraphs, $V_0 = 14.3\aleph$ (MeV).

The Quantic method of estimating a ground state is to rewrite p as $1/r = \mu/y$, to find the minimum of $E(y)$, and use that minimum point y_0 to define $E(y_0)$, which is to be interpreted as a qualitative estimate of the ground-state energy. The minimization statement is given by the vanishing of the derivative of (12.46) at $y = y_0$,

$$0 = -\frac{2}{y_0^3} + \frac{V_0}{M}y_0[y_0^2 - 4]e^{-\frac{y_0^2}{2}}, \tag{12.47}$$

and the customary way of solving such a problem is to solve (12.47) for y_0, and then substitute that value of y_0 back into (12.46) to obtain the binding energy. But since $\aleph$ is essentially unknown - one might guess it to be on the order of 1, representing a strong, nuclear force - and because we do want to represent the bound state energy as $E(y_0) = -2.2$ MeV, let us use that number together with the value of M to solve for y_0; and then solve for the value of $\aleph$.

To do this, combine (12.46) and (12.47) in such a manner that the $\exp[-y_0^2]$ factors of both equations are canceled, which produces

$$y_0^2(y_0^2 - 4)\left[1 + y_0^2\frac{|E|}{M}\right] = 2(y_0^2 - 4) \tag{12.48}$$

which is a cubic equation in y_0^2. From the graph of Figure 12.3, one sees that the minimum of the potential lies close to $y_0 = 2$, which suggests that the minimum of the energy should be somewhat larger; and this suggests the choice $y_0 = 3 + \Delta$ as a reasonable choice for the approximate solution of (12.48), retaining terms of no higher order than Δ (under the subsequently verified assumption that $\Delta^2 \ll 6|\Delta|$). This leads to the result: $\Delta \simeq .33$ and $y_0 \simeq 3.33$. Upon substituting this value of y_0 into (12.46) there follows $\aleph \simeq 1.25$, which provides the expected order of magnitude for a strong-coupling process. One may expect that when contributions from quark and nucleon spins are included, that number will change slightly, retaining its strong-coupling character.

12.5 Summary and Speculation

While the arguments put forth above are concerned with a realistic version of QCD, and have for simplicity neglected flavors, and electroweak interactions, quark and nucleon spin dependence, and have suppressed several "angular" integrations, all of which can be restored, as desired, the result is an explicit, model "deuteron" potential, of sufficiently short range and of the right order of magnitude to be considered as a qualitative derivation of nucleon-nucleon forces from basic realistic QCD.

Our tentative choice of $\gamma(E) = (\mu\delta)^4 = (\mu/E)^2$ has turned out to be qualitatively correct; and, in an eikonal context, this is interesting because it suggests that for the exchange of a composite object - in this case, the gluon bundles supporting a quark loop - between two "scalar" nucleons, the $\gamma(E)$ factor is not just what one typically finds when exchanging scalar quanta, $(m/E)^2$, but retains the memory and has a signature of the composite structure being exchanged, $\gamma(E) = (\mu/m)^2(m/E)^2$.

The above analysis should be almost immediately applicable to high-energy nucleon-nucleon scattering; and it will be interesting to see if the result of that calculation corresponds to the physical arguments recently suggested by [Islam (2011)].

Generalizations of this two-nucleon deuteron model to the construction of heavier, stable nuclei may well be possible, and might provide at least a partial basis for the nuclear shell model and the independent boson (IBM) model. In the first case, one would ask how many nucleon-generated gluon bundles can be attached to a single quark loop; and for the IBM model, asking how effective would attractive pairwise interactions of the deuteron form be when exchanged between nucleons in a three-dimensional array.

Finally, on a more fundamental level, it will be most interesting to see just how the structure of renormalization theory turns out for realistic QCD, the theory which has, built-in, quark transverse imprecision. From the experience gained in our work so far, the simplifications in which non-perturbative gluon exchanges organize themselves into gluon-bundle exchange displaying Effective Locality suggest that truly non-perturbative renormalization will turn out to be simpler than that of QED. One hopes to be able to answer this question in the near future.

PART 4
Astrophysical Speculations

Astrophysical Speculations

In Chapter 8, a new, symmetry-breaking, bootstrap solution for a QED-based vacuum energy (VE) corresponding to Dark Energy (DE), was formulated. That solution is based upon fluctuating lepton and quark pairs in the Quantum Vacuum (QV), which serve to define a VE of very high frequency, on the order of the Plank mass M_F. Gauge- and effective Lorentz-invariance are easily incorporated; and even a cursory glance at the form of that VE density, Figure 8.1 of that Chapter, suggests application to both Dark Energy and Inflation.

However, there is one crucial difference between the QED mechanism which can generate a present-day VE, associated with DE and that which is suitable for Inflation, describing how our Universe evolved from a speck of infinitesimally-small, space-time dimensions. That distinction appears because present-day lepton and quark pair fluctuations are described in terms of renormalized charge, as verified by the 27 megacycle contribution — out of a total of approximately 1057 — of the $2S_{1/2} - 2P_{1/2}$ Lamb Shift separation in Hydrogen. But in order for charge renormalization to exist, one must be able to view the charge at distances at least as large as the Compton wavelength of the particle carrying the charge; and this is impossible in the context of Inflation, where distances are less that $10^{-22\pm6}$ cm. In order for this QED VE model to be applicable here, one must resort to working within a formalism that contains only the unrenormalized charge, e_0. Is this possible?

In fact, it is quite possible, following the functional analysis of QED charge renormalization presented in Chapter 6, where the summation of the contributions of an infinite number of loop functionals L[A] suggested that charge renormalization is indeed finite; and to within the qualitative approximations of that estimate, indicate that the fine-structure constant calculated with e_0, rather than the renormalized e_R is given by $\alpha_0 = \pi/2$; and this will be the value used when writing the unrenormalized (but appropriately cut-off) $\Pi(k^2)$.

In order to formulate the QED-based model of Chapter 8 for Inflation, we are therefore committed to using α_0, rather than $\alpha \simeq 1/137$, in the unrenormalized contributions of the vacuum loops; and here a new difficulty arises, for the term corresponding to the unrenormalized $\Pi(k^2)$ is no longer real, but contains an imaginary part which increases as each lepton loop is added to the calculation. Remembering that the analysis of Chapter 8 leads to a VE density whose initial pulse seems destined to describe Inflation, one may ask if there is any simple and obvious way of removing those imaginary contributions to the unrenormalized $\Pi(k^2)$.

Again the answer is positive, and is suggested by the form of the solution for $\tilde{A}_\mu^{\text{vac}}$: If, in the QV, in addition to every lepton and quark loop fluctuation, there also exists a corresponding (massive, electrically-charged) fermionic tachyon loop fluctuation, it will contribute a negative imaginary term which exactly cancels that of the lepton or quark loop. This is a mathematical fact, and should be taken seriously in the sense of drawing a line between prejudice against any attempt of tachyon inclusion in modern Physics — a prejudice once strongly held by the author — and the possible, physical reality of such inclusion. As a way of retaining the QED-based model of VE in a form suitable to Inflation, all that is being assumed is the existence of massive, electrically charged $T - \bar{T}$ pairs in the QV; and to this there can be no experimental objection. Prejudice certainly, and the distaste of including a-causal objects in any theoretical explanation of any physical process; but such objections are not experimental.

In the following Chapters we shall see the minimal effects of tachyon *a*-causality, and note the irony of tachyon inclusion which, followed to its logical end, can provide a causal description of Dark Matter, of Inflation, of the Big Bang, and of the Birth and Death of a Universe.

Chapter 13

Inflation as the Precursor of Dark Energy

13.1 Introduction

For the present discussion of Inflation, describing how our Universe evolved from an origin of space-time dimensions infinitesimally small, it must be realized that the first step necessary in the description of the QED-based Model of Dark Energy of Chapter 8 — charge renormalization — suffers a certain disruption. This occurs because the process of renormalization requires the existence of spatial dimensions larger than the Compton wavelength of the charged particle defining the simplest vacuum fluctuation loop, or bubble. If such distances do not yet exist, renormalization is inconceivable, impossible, and hence irrelevant; and one must resort to working within a formalism which contains only the un-renormalized, or "bare" charge. Is this possible?

Happily, the answer is positive, for with the aid of intuitive and qualitative arguments, which extract the main UV logarithmic divergences from every perturbative term contributing to charge renormalization, it is possible to argue that the old idea [Johnson, et al. (1967)], [Baker and Johnson (1969)] of an eventual cancellation of all such divergences can realistically occur; and that the un-renormalized electric charge can be assigned a finite value [Fried (2012)]. All relevant details may be found in Chapter 6, where it was shown that, in contrast to the renormalized fine-structure constant $\alpha = e^2/4\pi \simeq 1/137$, the bare $\alpha_0 = e_0^2/4\pi = \pi/2$, where e_0 denotes the bare charge.

In its un-renormalized form, the $\Pi(k^2)$ of Chapter 6 may be written as

$$\Pi(k^2) = -\frac{2\alpha_0}{\pi} \int_0^1 dy \cdot y(1-y) \int_0^\infty \frac{ds}{s} e^{-is[m^2+y(1-y)k^2]}, \qquad (13.1)$$

where m denotes the mass of the charged particle whose vacuum loops are the source of the vacuum field, and where we shall replace $2\alpha_0/\pi$ by 1; the UV logarithmic divergence of (13.1) arising near $s = 0$ will require a physical cut-off, as appropriate. The dimensions of s are (inverse mass)^2, and the lower cut-off, ϵ, which needs to be inserted as the lower limit of the s-integral of (13.1), should presumably signify that distances smaller than the inverse Planck mass cannot be described within a physical theory that respects both Quantum mechanics (QM) and General Relativity (GR): $\epsilon \simeq (M_P)^{(-2)}$, with $M_P = 10^{(18)}$ GeV.

For the moment, let us be mathematical and replace the cut-off ϵ by δ, where δ is a parameter which may be allowed to vanish wherever possible without generating an UV logarithmic divergence. A simple change of contour is then useful, in which $\int_{\delta}^{\infty} ds$ is replaced by integration around a quarter-circle of radius δ , where $s \to \delta \exp(i\theta)$, from $\theta = 0$ to $\theta = -\pi'/2$, to which is added an integration along the negative imaginary axis, from $s = -i\delta$ to $s = -i\infty$. Upon setting $\delta = 0$, the contribution of the angular integration is simply $-i\pi/2$, while the variable change $s = -i\tau$ in the remaining integral produces the combination

$$\Pi(M^2) = \frac{i\pi}{12} - \int_0^{\prime} dy \cdot y(1-y) \int_{\delta}^{\infty} \frac{d\tau}{\tau} e^{-\tau[m^2 + y(1-y)M^2]}, \tag{13.2}$$

in which the second RHS term is clearly real, but the first term is imaginary, conflicting with the Model's demand that $\Pi(M^2)$ be real. Clearly, the inclusion of other lepton or quark loops on the RHS of (13.2) will only acerbate this situation, and we are lead to the Question: Is there any way of eliminating such unwanted, imaginary contributions, one that does not violate any principle of QFT?

Again the answer is positive, and it is suggested by the 'tachyonic' argument of the delta function of (8.11). Suppose that the quantum vacuum, in addition to containing virtual, charged lepton-antilepton loops also contains electrically-charged, tachyon-antitachyon loops, fermionic quantities which couple to photons in the same way as do leptons, but whose internal symmetries are somewhat different [Fried and Gabellini (2007)]. Suppose that the Universe contains this built-in symmetry, so that for every virtual lepton and quark loop fluctuating in the quantum vacuum there is a corresponding charged-tachyon pair also fluctuating in that vacuum. The fact that real, as opposed to virtual, tachyons would violate causality is irrelevant to this assumption, as is the fact that real leptons and hadrons can never be accelerated to velocities greater than c. We simply ask the reader to imagine that charged tachyon loops might exist in the quantum vacuum, one for every lepton and quark loop. There is nothing in experimental Physics which contradicts this assumption.

In what way could the existence of such virtual processes affect our physical world, which has never had the experimental need to consider tachyons as real particles? Such knowledge could appear as the result of an astrophysical catastrophe, such as a supernova explosion, in which strong electric fields acting for a short time in a restricted space could produce a Schwinger Mechanism that would, literally, tear such charged tachyon pairs out of the quantum vacuum, and send them, as real, massive, charged tachyons and anti-tachyons out into galactic space. It is the binding of real, charged tachyons by galactic magnetic fields that can represent the Dark Matter associated with every galaxy, as suggested in reference [Fried and Gabellini (2007)]. As described there, an electrically-charged, fermionic, high energy tachyon IS the perfect Dark Matter candidate, for although it can emit photons, *a la* Cerenkov, it almost immediately re-absorbs the photon it has just emitted.

If such massive, charged tachyons existed, their *a*-causal dynamics can be described in a manner quite similar to but different from causal leptons. The relevant statement here is that the corresponding, closed-tachyon-Ioop functional, $L_T[A]$ takes on the same form as that of the lepton L[A], except for the change of sign of its (mass)^2 term, where the $(m_\ell)^2$ of L[A] is changed to $-(m_T)^2$ inside $L_T[A]$.

(There is actually a small change in the effective tachyon-electromagnetic coupling due to the necessity of including a strong $T-\bar{T}$ binding interaction in the formalism, to prevent a photon from spontaneously decaying into a $T-\bar{T}$ pair. To prevent this, the value of that binding energy, B, must be larger than the tachyon mass. If that binding mechanism is modeled as a vector interaction, and if its effective coupling parameter is on the order of $M^{(-2)}$, then the order-of-magnitude calculation reported here will not be seriously changed, and this effect may be neglected.)

The tachyonic contribution of each such virtual pair, to be added to the lepton contribution of (13.2), is therefore

$$\Pi_T(M^2) = -\int_0^1 dy \cdot y(1-y) \int_\delta^\infty \frac{ds}{s}\, e^{+is[m_T^2 - y(1-y)M^2]}, \tag{13.3}$$

and a rotation of contour into the upper-half s-plane resembles that of (13.2), with the important difference that its imaginary term has the opposite sign to that of (13.2); and their sum vanishes, leaving the real contribution from this lepton-tachyon pair,

$$\Pi_{\ell+T}(M^2) = -\int_0^1 dy \cdot y(1-y) \int_\delta^\infty \frac{d\tau}{\tau}\Big[e^{-\tau[m_\ell^2+y(1-y)M^2]} + e^{-\tau[m_\tau^2-y(1-y)M^2]}\Big]. \tag{13.4}$$

In order to require that (13.4) remains real, we must require - in effect, we have already assumed - that $m_T > M/2$, for the maximum value of $y(1-y)$ in the range $1 > y > 0$ is $1/4$. We also expect that M will be much larger than any of the lepton (or quark) masses, which then disappear from the problem, while it will be convenient to set $m_T = \eta M$, where $\eta > 1/2$. Further, a simple variable change, and an integration by parts in both integrals of (13.4) generates the result

$$\Pi_{\ell+T}(M^2) = \int_0^1 dy \cdot y(1-y)\Big\{2C + \ln\left[\frac{m_\ell^2}{M_P^2} + y(1-y)\frac{M^2}{M_P^2}\right] + \ln\left[\frac{M^2}{M_P^2}(\eta^2 - y(1-y))\right]\Big\}, \tag{13.5}$$

where C is Euler's constant $\simeq .577$, and δ has been allowed to vanish whenever this can be done without generating a UV singularity. We have, in (13.5), replaced the mathematical δ by the physical cut-off $\epsilon = (M_P)^{(-2)}$, and now assume that the η-dependence is significantly larger than 1, so that the y-dependence inside the logarithms is irrelevant. To (13.5) we add the remaining two lepton-tachyon loop contributions, using the same, or an averaged value of η, to obtain

$$\sum_{\ell,T} \Pi_{\ell+T}(M^2) \to C - \ln\left(\frac{M_P^2}{\eta M^2}\right). \tag{13.6}$$

We neglect the insertion of quark loops and their corresponding tachyon partners, which lead to a small but unimportant change to what follows from using (13.6) as it stands.

13.2 Computation

When (13.6) is set equal to 1/2, as required by this QED-based, symmetry-breaking Model of Vacuum Energy, one obtains

$$\frac{M_P^2}{\eta M^2} \simeq e^{+.077} \simeq 0(1),$$

or $M_P \simeq \sqrt{\eta}M$, but $m_T = \sqrt{\eta}$, $M_P \geqslant M_P$. Here is an example where the tachyon mass may be larger than the Planck mass, a feature which will generate a most interesting Speculation in the following Section.

Even a cursory glance at Figure 8.1 suggests a possible connection with Inflation, and it is the leading peak of that energy density which we shall associate with Inflation, beginning near $x = 1$. Inflation is supposed to be in full swing during the growing-half of that first pulse, and decreases during the second half of that pulse, ending near $x = 4$. We shall simply choose the Liddle and Lyth [Liddle and Lyth (2000)] parameters for the initial time of inflation, t_i , and the value (in units of GeV) for the initial energy density ρ_i, parameters stated without any uncertainty; and then choose our two Model parameters to match the end time of Inflation, t_f associated with the average energy density at that time, ρ_f. If the initial parameters are allowed any reasonable variations, there could then be corresponding variations of the Model parameters.

Following reference [Liddle and Lyth (2000)], we therefore choose

$$t_i = \frac{1}{c}\frac{1}{M_P} \rightarrow \frac{1}{c}\left(\frac{\hbar}{M_P c}\right) \simeq 10^{-42} \text{ sec},$$

and from the Model write the expression for t_f in terms of η,

$$t_f = \frac{1}{c}R_f = \frac{x_f}{cM} \simeq \frac{4}{c}\left(\frac{\hbar}{Mc}\right) \simeq \frac{4}{c}\left(\frac{\hbar}{M_p c}\right)\sqrt{\eta} \text{ or}$$
$$t_f \simeq 10^{-42}\sqrt{\eta} \text{ sec.} \tag{13.7}$$

Setting this equal to the Liddle and Lyth value of $10^{(-32\pm6)}$ yields $\sqrt{\eta} = 10^{(10\pm6)}$, $M \sim 10^{(8\pm6)} GeV$, and $m_T \sim 10^{(28\pm6)} GeV$.

The only surprise here is the possibly large size of the 'average' tachyon mass, exceeding the Planck mass; but theoretical changes to the exponent of the expected t_f value could well decrease this value of m_T . But one may wonder if there is a cosmological significance to a tachyon mass far exceeding M_P.

Inspection of Figure 8.1 shows that the energy density in that first pulse is roughly $.02 \cdot \xi \cdot M^4$, or $p_f^{\frac{1}{4}} \sim 10^{\frac{-1}{2}} \cdot \xi^{\frac{1}{4}} M \sim 10^{8\pm6} GeV$, which lies within the bounds of the listed $10^{(13\pm3)}$ if one chooses $\xi^{\frac{1}{4}} \sim 0(1)$. In this way, and with (to within an order of magnitude) the same parameter as needed for the Dark Energy estimation, this QED vacuum energy Model can satisfy the cosmological requirements for both Inflation and Dark Energy.

13.3 A Cosmological Speculation

From the above paragraphs, one sees that the tachyon mass can, in general, be larger than the Planck mass, and it is amusing to follow this line of thought using the above estimates as a guide. We shall still adhere to the concept that the smallest coordinate differences compatible with GR and QM are given (in appropriate units) by the inverse of the Planck Mass. But what we shall imagine here is that the initial vacuum energy deposited as the new Universe appears is not the Planck mass, with an initial energy density of $(M_P)^4$; but rather that this happy event was generated by the annihilation of a random $T - \bar{T}$ pair with total energy of perhaps $10 m_T$ (to be conservative).

Then the $t_{i,f}$ and $R_{i,f}$ calculations go through as before, because those estimates were based on the inverse Planck length, to which we continue to adhere as the smallest possible coordinate difference. But at the end of Inflation, one would now find $\rho_f \sim (10\ m_T)/(\frac{4\pi}{3})R_f^3$, or $\rho_f^{1/4} \sim 10^{(15\pm3)}$ GeV, which is very close to the value quoted in Liddle and Lyth. To the author, this seems like a rather more elegant way of fitting data, and it may perhaps suggest a somewhat daring Cosmological Speculation, as follows.

Suppose that a sizable amount of energy, much larger than M_P, is suddenly deposited — for example, by the accidental annihilation of a highly energetic $T - \bar{T}$ pair — at one point in a coordinate system whose space-time structure cannot support that much energy. What may well happen at that point is that a new coordinate system, of a New Universe, appears with no memory of its origin, and a corresponding Inflation begins. At that point, or in the extremely small region at which this occurs, one may imagine that the Old Universe's space-time structure has been "torn", or disrupted in such a way that - in that region only - the separation of vacuum energy from real energy is disrupted, and that the immense amount of the Old Universe's vacuum energy is able to force its defining vacuum fluctuations through that tear, and in the process convert them to real lepton-anti-lepton, real quark-anti-quark, and real tachyon-anti-tachyon pairs. In brief, this is the Big Bang of the New Universe.

One can also think of this as a spectacular Schwinger Mechanism, wherein the potential energy of the quantum vacuum is able — at that point — to convert to "real" energy, and then tear the vacuum fluctuations, which originally defined that energy, out of the Old Universe's vacuum energy. The details, of course, can only be imagined; even for terrestrial events, it is difficult to describe an explosion in terms of probabilistic effects. But one may note that this vision of a New Universe's Inflation and its Big Bang will satisfy conservation of energy and electrical charge.

Finally, it is difficult to refrain from considering the fate of the Old Universe, if that entity continues to lose a sizable portion of its vacuum energy to the New Universe. For it is, in this Model of Inflation and Dark Energy, the potential energy locked in the quantum vacuum which serves to resist the mutual gravitational attraction of a universe's many parts; and if this vacuum energy, or a too-large portion of it, is lost to the New Universe, the result must be the collapse of the Old Universe into a monstrous black hole, one whose radiation could well be observable astronomically, by astronomers of the New Universe.

13.4 Summary

We remind the reader that the above paragraphs come at the end of a line of intuitive reasoning, which began with a symmetry-breaking Model of a possible QED vacuum energy. Tachyons were introduced only when demanded by the Model, in order to remove an unwanted, imaginary contribution. Perhaps one might also note that the author's prejudice against tachyons has been continuous and of long standing; but only in recent years has it become clear that such a prejudice, which appears to stand in the way of understanding astrophysical data, must be discarded.

Chapter 14

Quantum Tachyon Dynamics

Because the possible existence of massive, charged, fermionic tachyons is central to the QED-based model of the previous Chapter, it is appropriate to end these astrophysical discussions with an elaboration of those properties which are peculiar to tachyons, and specifically, those of the fermionic persuasion. While not relevant to everyday terrestrial considerations, such entities have the possibility of contributing to, or even being the sole explanation of Dark Matter; and can also provide an explanation of gamma-ray bursts with and without x-ray and optical tails, as well as ultra-high-energy cosmic rays that violate the GZK limit. Even the recent "Fermi Bubbles" visible above and below our galactic plane, have a partial explanation in terms of a specific astrophysical conglomeration of such tachyons. And, as suggested in the last Section of the previous Chapter, a random annihilation of a $T - \bar{T}$ pair of energy much larger than the Planck mass may be the event that sparks the creation of a new Universe.

The present Chapter defines a quantum field theory of tachyons, particles similar to ordinary leptons, but with momenta larger than energy. The theory is invariant under the full CPT transformation, but separately violates P and T invariance. Micro-causality is broken for space-time intervals smaller than the inverse tachyon mass, but is effectively preserved for larger separations. Electrically-charged fermionic rather than charged scalar tachyons are considered, in order to maximize the probability of the reabsorption of Cerenkov-like radiation by the tachyon, thereby permitting a high-energy tachyon to retain its energy over galactic distances; and in so doing, acting as a 'dark particle'.

Topics treated in the following Sections include the choice and Schwinger Action Principle variations of an appropriate Lagrangian, spinorial wave functions, relevant Green's functions, a description of an S-Matrix and

Generating Functional, and a variety of interesting kinematical processes including photon emission and reabsorption, and relevant annihilation and scattering effects. A version of Ehrenfest's Theorem is developed in order to provide a foundation for a classical description of charged tachyons in an external electromagnetic field. Applications are then made to several outstanding astrophysical puzzles: dark matter, gamma-ray bursts, ultra-high-energy cosmic rays, and the "Fermi bubbles" recently observed above and below our Galaxy. Much of this material has been taken from the paper by the author and Y. Gabellini [Fried and Gabellini (2007)].

14.1 Introduction

Two immediate theoretical observations can be made to set the stage for this discussion. The Lagrangians typically used to define the Higgs mechanism, and in particular the Weinberg–Salam method of defining a "true" ground state, are basically theories of "condensed" tachyons, theories which use a negative $(\text{mass})^2$ and a self interaction to break spontaneously the initial vacuum symmetry, and arrive at a minimum energy ground state. But if a strong, external, electromagnetic field is locally present, and if the quanta of the tachyon field are charged, that ground state need not be, locally, the correct vacuum state; and one might expect fluctuations about that W-S vacuum. Those fluctuations could well be tachyonic, as expected from the form of the Lagrangian used, suggesting that pairs of charged tachyons could, in principle and just as for ordinary leptonic pairs, be produced by extremely strong electromagnetic fields.

Secondly, consider the non-perturbative Schwinger mechanism [Schwinger (1951)] for the production of lepton pairs by a strong electric field, a calculation which has received during the past half century a great amount of attention and generalization[1]. Physically, the vacuum's charged lepton pairs which continuously fluctuate into and out of existence are acted upon by the continuous distribution of virtual photons comprising the external electric field; and if and when sufficient amounts of energy E and momentum p have been transferred to each member of a pair, they become real leptons, each with $E > p$, Suppose now that charged, tachyonic degrees of freedom exist, perhaps the same ones whose "condensed"

[1] See, for example, [Hounkounnou (2000)]. More recent attempts at incorporating arbitrarily varying electromagnetic fields can be found in the papers [Tomaras (2000)], [Fried and Woodard (2002)] and [Avan, et al. (2003)].

summation defines the spontaneously broken vacuum. Then it is entirely reasonable to imagine that extremely strong electric fields could transfer sufficient energy and momentum to the tachyon pair, with each member here displaying $p > E$. There is no conservation law that forbids this; in principle, it is possible.

Of course, one immediate objection to $v > c$ motion is a resulting lack of micro-causality. But if the tachyon mass M which sets the scale for acausal effects is sufficiently large, e.g., $M \gtrsim 10^7$ GeV, this difficulty will not be particularly relevant at typical laboratory distances. Nevertheless, the actions of charged tachyons after they are produced may well be unusual, especially at galactic distances.

Mention should be made of the Wigner classification [Wigner (1939)], [Beckers (1978)] of relativistic particles, wherein it is well-known that a scalar tachyon field need have but one component, but that the Little Group of "higher dimensional" tachyons is given by the non-trivial representations of SU (1,1), which has only, one hermitian generator leading to a unitary representation in finite dimension, instead of the three for conventional massive particles for which the spin can be defined and measured at rest. In other words, for a spin 1/2 massive particle, the generators of the Little Group $SU(2)$ are $\{\sigma_x, \sigma_y, \sigma_z\}$, and the spin of the particle can be measured in any direction. On the other hand, for a two component tachyon (for which no spin exists), the generators are $\{i\sigma_x, i\sigma_y, \sigma_z\}$, and only one direction — z — leads to a measurable quantity. This loss of rotational invariance will be manifest in the next paragraph. Nevertheless, our model field theory represents its leptonic tachyons in terms of four components, although somewhat modified from those of Dirac; and the "spin" will be discussed in Appendix G. We also insist on the restriction to orthochronous Lorentz transformations (LTs): so that — in contrast to the work of [Dhar (1968)] — we absolutely forbid LTs which can interchange positive and negative energy tachyons. An amount of causality sufficient to define scattering states requires a special assumption, but can be arranged. Invariance under the full CPT operation is maintained, although P and T invariance are separately violated. And we stress that the micro-causality of this model field theory is broken at very small space-time intervals. But if massive, charged tachyons provide a partial explanation of how the large scale universe works, so be it.

Before writing any equations, we remind the reader of the Minkowski metric that we are using: $a_\mu = (\vec{a}, ia_0)$, so that $a^2 = \vec{a}^2 - a_0^2$. Since the equations of General Relativity do not appear in this Chapter, there is no distinction between covariant and contravariant indices. Our gamma matrix notation is defined immediately after Equation (14.2).

Several decades ago, other authors [Dhar (1968)] have discussed the possibility of zero charge tachyons, and produced somewhat simpler models than the leptonic tachyons suggested here. Following this spirit, it would he possible to describe spin-zero charged tachyons, which couple in a standard, gauge invariant manner to the photon field, with an interaction term of the form:

$$ieA_\mu(\varphi^\dagger\partial_\mu\varphi - \partial_\mu\varphi^\dagger\varphi) + e^2\varphi^\dagger A^2\varphi \tag{14.1}$$

where A_μ denotes a conventional photon field operator. What makes this theory unsuitable for our astrophysical purposes is the $e^2\varphi^\dagger A^2\varphi$ term of this interaction, which can describe the simultaneous emission of a pair of photons; in this case the Cerenkov-like angular restrictions of single photon emission and absorption are absent, which restrictions are present when but a single photon is emitted by a tachyon. Crucial to the astrophysical usefulness of the present model tachyons is the strong probability that a very high energy tachyon can reabsorb every photon that it emits, and thus retains its high energy across astrophysical times and distances. But if two photons were emitted simultaneously, the probability of reabsorption of the photon moving in a direction roughly opposite to that of the tachyon would be small.

The observant reader will shortly notice that our subsequent remarks are relevant to the interaction of such tachyons with an electromagnetic field specified by a vector potential, which can represent either an external, classical field $A_\mu^{\text{ext}}(x)$, or the photon operator $A_\mu(x)$. No reference will be made to a possible non-linear coupling. e.g., $\mathcal{L}'' = -(\lambda^2/2)(\bar{\psi}_T\gamma_5\psi_T)^2$, which would be appropriate in a Higgs or W-S symmetry breaking context (in which one neglects the gradient portions of the Lagrangian, as well as any electromagnetic coupling). But there is one electromagnetic situation in which such a $\mathcal{L}''$ might not be neglected, for a way must be found to block the decay of a mass shell photon into a pair of mass shell tachyons. For ordinary, mass shell particles, the reaction $k \to p + \bar{p}$ is kinematically impossible, whereas $k \to T + \bar{T}$ is, kinematically, perfectly possible. Were this reaction not forbidden, perhaps by a selection rule, or severely damped, or given an energy threshold below which it is forbidden, any real photon of any energy would be able to decay into a $T\bar{T}$ pair, and our world would be quite different.

From a dynamical point of view, there are two possible sources of the needed damping, or threshold effect. One is the realization that the vertex $k \to T+\bar{T}$ will resemble tachyon scattering by a photon, except for the huge "momentum transfer" needed to replace one T leg by a $\bar{T}$ leg; and, as such, serious, soft photon damping of this vertex may be expected. The second source is damping induced by $\mathcal{L}''$, which, at least in eikonal approximation, can be extracted from the radiative corrections that it generates.

To make this argument more explicit, consider the set of Feynman graphs entering into the amplitude $k \to T + \bar{T}$, which include the interactions generated by $\mathcal{L}''$ along each T and $\bar{T}$ leg, and, in particular, between T and $\bar{T}$. In effect, $\mathcal{L}''$ represents a λ-strength attraction between T and $\bar{T}$, which must be overcome if the T and $\bar{T}$ are to "unbind". The statement that it is difficult to break them apart is, quantum mechanically, the statement that unless k_0 is extremely large, the amplitude for this process is very small; and just this effect is suggested by the (non perturbative) eikonal approximation mentioned above. Even more striking is a simple, cluster expansion approximation of such radiative corrections to the Schwinger mechanism, which suggests that the electric field needed for $T + \bar{T}$ production has an effective, λ dependent threshold. In Section 14.5.1, a simple calculation suggests how the production of a $T\bar{T}$ pair by a real photon can be forbidden, due to the presence of $\mathcal{L}''$. In this Chapter, we shall only be concerned with the electromagnetic interactions of such tachyons after they are produced.

The arrangements of these remarks is as follows. In the next section, a "misgenerated lepton" model of tachyons is defined, and its spinorial content outlined. In Section 14.3, a complete, functional Quantum Tachyon Dynamics (QTD) is set up, with coupling to the electromagnetic field, in analogy to QED. In the next Section, the possibility of photon emission and reabsorption by charged tachyons is discussed, and in Section 14.5 the kinematics of tachyon particle reactions is outlined. A version of Ehrenfest's theorem is exhibited in Section 14.6, to set the stage for a subsequent "Classical Tachyon Dynamics", describing the classical motion of a charged tachyon in an external electromagnetic field. A final Section suggesting an explanation of recently measured "Fermi Bubbles" and Appendices F and G completes the presentation.

14.2 QTD as a QFT

The QFT here called QTD is modeled as closely as possible upon QED, with but two changes, and one change of interpretation. The standard QED Lagrangian may be written

$$\mathcal{L} = -\bar{\psi}[m + \gamma \cdot (\partial - ieA)]\psi + \mathcal{L}_0(A) \tag{14.2}$$

where $\mathcal{L}_0(A)$ denotes the free photon Lagrangian, and m and e are the electron bare mass and charge, respectively; in all of the following, hermitian γ_μ satisfying $\{\gamma_\mu, \gamma_\nu\} = 2\delta_{\mu\nu}$ will be used, with $\{\gamma_5, \gamma_\mu\} = 0$, $\gamma_j = -i\gamma_4\alpha_j$, $\alpha_j = \sigma_1 \bigotimes \sigma_j$, $\gamma_5 = \gamma_1\gamma_2\gamma_3\gamma_4$, where the elements of $\gamma_{4,5}$ are two-by-two matrices: $\gamma_4 = \begin{pmatrix} 1 & 0 \\ 0 & -1 \end{pmatrix}$, $\gamma_5 = -\begin{pmatrix} 0 & 1 \\ 1 & 0 \end{pmatrix}$. The σ_j denote the hermitian Pauli matrices, satisfying $\sigma_i\sigma_j = \delta_{ij} + i\epsilon_{ijk}\sigma_k$. We use the Minkowski metric: $x_\mu = (\vec{x}, ict)$, with $\hbar = c = 1$ and $\bar{\psi} = \psi^\dagger\gamma_4$.

From (14.2) and the Schwinger Action Principle, one obtains both the field equations:

$$[m + \gamma \cdot (\partial - ieA)]\psi = 0 \tag{14.3}$$

and, from the surface-term variations of the Action integral of (14.2), the equal time anti-commutation relation (ETAR):

$$\{\psi_\alpha(x), \bar{\psi}_\beta(y)\}_{x_0 = y_0} = \gamma_4^{\alpha\beta}\delta(\vec{x} - \vec{y}) \tag{14.4}$$

with (14.4) providing a statement of micro-causality, as expected in a causal theory.

The first, and simplest, departure from QED that shall be adopted here is the replacement of m by iM, where M is understood to be large, $M \gg m$:

$$\mathcal{L} = -\bar{\psi}_T[iM + \gamma \cdot (\partial - ieA)]\psi_T + \mathcal{L}_0(A). \tag{14.5}$$

This is not the final tachyon Lagrangian, but only the simplest, initial form written to display the needed $m^2 \to -M^2$ continuation. For simplicity, we allow the tachyon field to be coupled to the electromagnetic field with the QED coupling constant, and first consider the properties of the free tachyonic field, $\psi_T^{(0)}$, satisfying:

$$(iM + \gamma \cdot \partial)\psi_T^{(0)} = 0 = \bar{\psi}_T^{(0)}(iM + \gamma \cdot \overleftarrow{\partial}). \tag{14.6}$$

One immediately sees that it will not be possible to build a conserved current from $\psi_T^{(0)}$ and $\bar{\psi}_T^{(0)}$ obeying (14.6).

In direct imitation of the free lepton fields' representations in term of creation and destruction operators, and spinorial wave functions, we write:

$$\psi_T^{(0)}(x) = (2\pi)^{-3/2} \sum_{s=1}^{2} \int d^3p \left[\frac{iM}{E(p)}\right]^{1/2} \{b_s(\vec{p})e^{ip\cdot x}u_s(p) + d_s^\dagger(\vec{p})e^{-ip\cdot x}v_s(p)\} \tag{14.7}$$

where $b_s(\vec{p})$ and $d_s^\dagger(\vec{p})$ are the destruction and creation operators, respectively, for T and $\bar{T}$, while $u_s(p)$ and $v_s(p)$ are their appropriate spinors; and $p \cdot x = \vec{p} \cdot \vec{x} - E(p)x_0$, with $E(p) = \sqrt{\vec{p}^2 - M^2}$. Perhaps the simplest description appears if $b_s(p)$ and $d_s(p)$ satisfy exactly the same anticommutation rules as for ordinary leptons. One then realizes that the $\exp(\pm ix_0\sqrt{\vec{p}^2 - M^2})$ factors of (14.7) will generate exponentially growing or damping time dependence of the $\psi_T^{(0)}$, $\bar{\psi}_T^{(0)}$, which is simply not acceptable. The obvious way to prevent this is to restrict $|\vec{p}|$ values to be larger than M, a solution which will have other consequences, but ones that are acceptable and reasonable in a non-causal theory. Henceforth, a factor of $\theta(|\vec{p}| - M)$ will appear under the momentum integrals of $\psi_T^{(0)}$ and $\bar{\psi}_T^{(0)}$.

Consider now the momentum-space equations which the tachyon T and anti-tachyon $\bar{T}$ must satisfy:

$$(M + \gamma \cdot p)u_s(p) = 0, \quad (M - \gamma \cdot p)v_s(p) = 0. \tag{14.8}$$

Solutions of (14.8) may be expressed in terms of the projection operators $\Lambda_\pm(p) = \frac{M \pm \gamma \cdot p}{2M}$:

$$u_s(p) = \sqrt{\frac{2M}{|\vec{p}|}}\Lambda_-(p)\xi_s, \quad v_s(p) = \sqrt{\frac{2M}{|\vec{p}|}}\Lambda_+(p)\xi_{s+2} \tag{14.9}$$

where, as for leptons, $s = 1$ or 2, and $\sqrt{\frac{2M}{|\vec{p}|}}$ is an appropriate normalization constant. The adjoint spinors $\bar{u} = u^\dagger\gamma_4$, $\bar{v} = v^\dagger\gamma_4$, are then given by:

$$\bar{u}_s(p) = \sqrt{\frac{2M}{|\vec{p}|}}\bar{\xi}_s\Lambda_+(p), \quad \bar{v}_s(p) = \sqrt{\frac{2M}{|\vec{p}|}}\bar{\xi}_{s+2}\Lambda_-(p) \tag{14.10}$$

and one sees that the norms of these states have the unfortunate property of vanishing: $\bar{u}_s(p)u_s(p) = \bar{v}_s(p)v_s(p) = 0$.

Does this signify that it is impossible to construct lepton-style tachyonic states? Not necessarily, because — as Dirac pointed out, long ago [Dirac (1943)][2]— one can define expectation values with the aid of an indefinite metric operator, $\eta = \eta^\dagger$, such that the proper norm of a state is given by $\bar{u}\eta u$, rather than $\bar{u}u$. In our case, there exists a clear candidate for this metric operator: $\eta = \gamma_5$, which has the pleasant property of converting $\Lambda_\pm$ to $\Lambda_\mp$ as it passes through either operator. However, there is a price to pay for this convenience, in that such norms are no longer scalars but pseudoscalars; and the vectors constructed from them will be pseudovectors, etc. It is difficult to avoid the suspicion that such tachyons may turn out to be associated with the electroweak interactions, forming the "condensate" of the W-S vacuum, although that subject will not be treated in this analysis.

Following the above discussion, we shall insist that a factor of γ_5 be inserted immediately after each $\bar{u}_s$ or $\bar{v}_s$ factor, for all matrix elements subsequently calculated. With this intention, the tachyonic part of the Lagrangian of (14.5) shall now be written as:

$$\mathcal{L}_T = -\bar{\psi}_T \, \gamma_5[iM + \gamma \cdot (\partial - ieA)]\psi_T \tag{14.11}$$

with field equations:

$$[iM + \gamma \cdot (\partial - ieA)]\psi_T = 0 = \bar{\psi}_T\gamma_5[iM - \gamma \cdot (\overleftarrow{\partial} + ieA)] \tag{14.12}$$

so that the γ_5 insertion is automatic. Were the spinors ξ_s chosen as for the leptons, $\xi_s = \chi_s$, where χ_1 is the column matrix of elements (1, 0, 0, 0), χ_2 has elements (0, 1, 0, 0), etc, then one easily calculates:

$$\bar{u}_{s'}(p)\gamma_5 u_s(p) = i\chi^\dagger_{s'}\vec{\sigma}\cdot\hat{p}\chi_s = i(\vec{\sigma}\cdot\hat{p})_{s's} \;\text{ with }\; \hat{p} = \vec{p}/|\vec{p}| \tag{14.13}$$

which, while non-zero, does not display the usual $\delta_{ss'}$, but corresponds to a linear combination of the spin states labeled by s and s'. Similarly one has $\bar{v}_{s'}(p)\gamma_5 v_s(p) = -i(\vec{\sigma}\cdot\hat{p})_{s's}$.

The closure, or completeness relation for such spinors is then given by:

$$\sum_{s',s=1}^{2} \{u^\alpha_s(p)(\bar{u}_{s'}(p)\gamma_5)^\beta - v^\alpha_s(p)(\bar{v}_{s'}(p)\gamma_5)^\beta\}(\vec{\sigma}\cdot\hat{p})_{s's} = \delta^{\alpha\beta}. \tag{14.14}$$

[2] A discussion of the indefinite metric can be found in the book by W. Heitler, "The Quantum Theory of Radiation" [Heitler (1954)].

The free field spinors now satisfy:

$$(M + \gamma \cdot p)u_s(p) = (M - \gamma \cdot p)v_s(p) = 0$$

and:

$$\bar{u}_s(p)\gamma_5(M + \gamma \cdot p) = \bar{v}_s(p)\gamma_5(M - \gamma \cdot p) = 0.$$

It should be noted that the Action formed from the Lagrangian of (14.11) is hermitian, as is the Hamiltonian. These assertions require the customary neglect of spatial surface terms, as well as the observation — obtain from the field equations — that: $\frac{\partial}{\partial t}\int d^3x\bar{\psi}_T\gamma_5\gamma_4\psi_T = 0$.

If the coefficient of A_μ in (14.11) is defined as the charged tachyon current operator: $J_\mu^T = ie\bar{\psi}_T\ \gamma_5\gamma_\mu\psi_T$, then it follows from the field equations (14.12) that this charged current is conserved: $\partial_\mu J_\mu^T = 0$. Charged conjugation may be effected by the same set of Dirac matrices used for the lepton case: $\psi_c = C\bar{\psi}^T$, where the superscript T denotes transposition and $C = i\gamma_2\gamma_4$.

Finally, one must discuss the equal-time anticommutation relations $\{\psi_T^\alpha(x), \bar{\psi}_T^\beta(y)\}_{x_0=y_0}$, which — as for ordinary leptons — are assumed to be the same for free and interacting field operators. The Schwinger Action Principle[3] underlying QFT specifies the ETAR by identifying the infinitesimal generators from the surface terms of the variation of the Action: $\delta W = -\int d\sigma_\mu\bar{\psi}\gamma_\mu\delta\psi$ in QED, or: $\delta W = -\int d\sigma_\mu\bar{\psi}_T\gamma_5\gamma_\mu\delta\psi_T$ for the present case of QTD. For the usual case where the relevant space-time surface is taken as a flat time cut, one has in QED:

$$\delta W = -\int d\sigma_4\bar{\psi}\gamma_\mu\delta\psi = i\int d^3y\bar{\psi}(\vec{y}, t)\gamma_\mu\delta\psi(\vec{y}, t)$$

thereby identifying the field momentum operator conjugate to ψ_α as $\pi_\alpha(\vec{y}, t) = (\bar{\psi}(\vec{y}, t)\gamma_4)_\alpha$ and from this follows the choice of ETAR: $\{\psi_\alpha(x), \psi_\beta^\dagger(y)\}_{x_0=y_0} = \delta_{\alpha,\beta}\delta(\vec{x} - \vec{y})$, which expresses the fermion microcausality of conventional QED.

For an acausal theory one should not expect to require strict microcausality; but, keeping to the original statement of the Action Principle, it is possible to obtain a modified expression of the QTD ETAR. which can be written in the form:

---

[3]See, for example, [Fried (1972)] as well as the original [Schwinger (1951)].

$$\{\psi_T^\alpha(x), (\bar{\psi}_T(y)\gamma_5)^\beta\}_{x_0=y_0} = (\gamma_4)^{\alpha\beta}\hat{\delta}(\vec{x}-\vec{y}). \tag{14.15}$$

It is then straightforward to employ (14.14) and show that the substitution of (14.7) and its adjoint into the LHS of (14.15) yields:

$$\hat{\delta}(\vec{x}-\vec{y}) = \int \frac{d^3p}{(2\pi)^3}\theta(|\vec{p}|-M)e^{i\vec{p}\cdot(\vec{x}-\vec{y})} \tag{14.16}$$

which is just the form one expects with the restriction $|\vec{p}| > M$. This "modified delta function" of (14.16) is a well defined distribution of $|\vec{x}-\vec{y}|$, which can be written in various forms, such as:

$$\hat{\delta}(\vec{x}-\vec{y}) = \delta(\vec{x}-\vec{y}) - \int \frac{d^3p}{(2\pi)^3}\theta(M-|\vec{p}|)e^{i\vec{p}\cdot(\vec{x}-\vec{y})} \tag{14.17}$$

where the second RHS of (14.17) is a finite measure of the lack of micro-causality of this tachyon theory. With $\vec{z} = |\vec{x}-\vec{y}|$, it is immediately evaluated as:

$$\frac{M}{2\pi^2 z^2}\left\{\frac{\sin(M\ z)}{M\ z} - \cos(M\ z)\right\}. \tag{14.18}$$

For small and large $M\ z$, (14.18) reduces to $\frac{1}{6\pi^2}M^3$, $M\ z \ll 1$; and to $-\frac{M}{2\pi^2 z^2}\cos(M\ z)$, $M\ z \gg 1$. In other words, the lack of micro-causality occurs mainly for extremely small distances, $z < M^{-1}$, where it is significant, while, for $z > M^{-1}$, the effect is oscillatory and of unimportant magnitude. The basic theory, however, is non-causal in a qualitatively different sense, with tachyon propagation emphasized outside the light cone but damped away inside, in exactly the opposite sense to that of QED.

The free tachyon propagator, defined as:

$$\begin{aligned} S_T^{\alpha\beta}(x-y) &= i < \left(\psi_T^\alpha(x)(\bar{\psi}_T(y)\gamma_5)^\beta\right)_+ > \\ &\equiv i\theta(x_0-y_0) < \psi_T^\alpha(x)(\bar{\psi}_T(y)\gamma_5)^\beta > -i\theta(y_0-x_0) < (\bar{\psi}_T(y)\gamma_5)^\beta\psi_T^\alpha(x) > \end{aligned} \tag{14.19}$$

is then, with (14.15), to satisfy:

$$\begin{aligned} (iM+\gamma\cdot\partial)S_T(x-y) &= \delta(x_0-y_0) < \{(\gamma_4\psi_T(x))^\alpha, (\bar{\psi}_T(y)\gamma_5)^\beta\} > \\ &= \delta(x_0-y_0)\hat{\delta}(x-y) \end{aligned}$$

and has the integral representation:

$$S_T(z) = -i \int \frac{d^4p}{(4\pi)^4} \theta(|\vec{p}| - M) \frac{e^{ip\cdot z}}{(M + \gamma \cdot p)}. \tag{14.20}$$

An immediate question is the method of choosing contours to perform the momentum integrals of (14.20), since: $(M + \gamma \cdot p)^{-1} = (M - \gamma \cdot p) \cdot (M^2 - p^2)^{-1}$. If, in the usual way, $M^2 \to M^2 \pm i\epsilon$, one must decide which possibility is appropriate. Calculations for the $\int dp_0$ yields:

$$S_T(z) = (\mp)\frac{1}{2} \int \frac{d^3p}{(2\pi)^3} \frac{\theta(|\vec{p}| - M)}{\sqrt{\vec{p}^2 - M^2}} e^{i\vec{p}\cdot\vec{z} \mp i|z_0|\sqrt{\vec{p}^2 - M^2}} (M - \vec{\gamma} \cdot \vec{p} \pm i\gamma_4 \epsilon(z_0)) \tag{14.21}$$

and if the $z \to 0$ limit of (14.21) is compared with a direct computation of the second line of (14.19), using the completeness statement (14.14), the two calculations will agree if the upper sign is used in (14.21), that is: $M^2 \to M^2 + i\epsilon$. For $z \neq 0$, one then finds agreement with the form of the fermion propagator in QED; in the sense that an exponential factor $\exp(i\vec{p}\cdot\vec{z} - i\omega|z_0|)$ is present in both theories, with $\omega = \sqrt{\vec{p}^2 + m^2}$ for QED and $\omega = \sqrt{\vec{p}^2 - M^2}$ for QTD. Henceforth, $M^2 \to M^2 + i\epsilon$.

Because of the $\theta(|\vec{p}| - M)$ factor, one might guess that the propagator of (14.20) and (14.21) is not Lorentz invariant (LI), but this turns out not to be the case; in fact, the propagator is LI, and is related to the conventional particle propagator by the simple continuation: $m \to iM$. The easy way to see this is to set $S_T(z) = (iM - \gamma \cdot \partial)\Delta_T(z, M)$, in analogy to the ordinary lepton case, where $S_c(z) = (m - \gamma \cdot \partial)\Delta_c(z, m)$; and to insert a factor of unity: $1 = \theta(\vec{p}^2 + m^2)$, under the integrals that define $\Delta_c(z, m)$. If the continuation $m \to iM$ is now made, one finds that:

$$\Delta_c(z, iM) = \Delta_T(z, M) = -\int \frac{d^4p}{(2\pi)^4} \theta(|\vec{p}| - M) \frac{e^{ip\cdot z}}{p^2 - M^2 - ie} \tag{14.22}$$

a completely LI result in which the space-like and time-like regions of $\Delta_c(z, m)$ have been interchanged. Explicit calculation shows that the integral representation of (14.22) does indeed generate precisely $\Delta_c(z, iM)$.

14.3 Functional QTD

In this section, a generating functional for QTD is constructed, in analogy with that of ordinary QFT, with charged tachyons replacing charged leptons. Because the properties of the free tachyon creation and destruction operators have been chosen to be the same as those of ordinary lepton fields, the transition from n-point functions of QTD to corresponding S-Matrix elements can he taken over directly from ordinary QED. However, because of the ubiquitous $\theta(|\vec{p}|-M)$ factors, one requires the understanding that tachyons IN and OUT field operators must be defined as "asymptotic times", that are outside the light cone of any origin of coordinates in terms of which the field operators might be evaluated.

For the Lagrangian:

$$\mathcal{L} = -\bar{\psi}_Y \gamma_5 [iM + \gamma \cdot (\partial - ie[A + A^{\text{ext}}])] \psi_T + \mathcal{L}_0(A) \tag{14.23}$$

where A denotes the ordinary, quantized, photon field, and $\mathcal{L}_0(A)$ is its free Lagrangian (in an appropriate gauge), one defines the generating functional in terms of tachyonic and photon sources as:

$$< S[A^{\text{ext}}] > \mathcal{Z}[\eta, \bar{\eta}, j] =< S[A^{\text{ext}}] \left(e^{i \int [\bar{\eta}\psi_T + \bar{\psi}_T \eta + j_\mu A_\mu]} \right)_+ > \tag{14.24}$$

just as one would do for QED. With the aid of the field equations, the photon ETCR and the tachyonic ETAR — or, directly, from the Schwinger Action Principle — one obtains [Fried (1972)]:

$$\begin{aligned} \mathcal{Z}[\eta, \bar{\eta}, j] = e^{\frac{i}{2} \int j D_c j} \cdot e^{-\frac{i}{2} \int \frac{\delta}{\delta A} D_c \frac{\delta}{\delta A}} \cdot e^{i \int \bar{\eta} G_T [A + A^{\text{ext}}] \gamma_5 \eta} \\ \cdot e^{L[A + A^{\text{ext}}]} / < S[A^{\text{ext}}] > \end{aligned} \tag{14.25}$$

where, for (14.25), $A_\mu = \int d^4 y D_{c,\mu\nu}(x-y) j_\nu(y)$. Because of the normalization requirement $\mathcal{Z}[0,0,0] = 1$ there follows:

$$< S[A^{\text{ext}}] >= e^{-\frac{i}{2} \int \frac{\delta}{\delta A} D_c \frac{\delta}{\delta A}} \cdot e^{L[A + A^{\text{ext}}]} |_{A \to 0} \tag{14.26}$$

where:

$$L[A] = i \int_0^e de' Tr[\gamma \cdot A \, G_T[e'A]] \tag{14.27}$$

and:

$$G_T[eA] = S_T[1 - ie(\gamma \cdot A) S_T]^{-1}. \tag{14.28}$$

Because the γ_5 factors always appear with a corresponding $\bar{\psi}_T$ in (14.15) and (14.25), there is no γ_5 presence in the log of the tachyon determinant of (14.27), nor in the Green's function of (14.28). This means that the mixing of vector and axial vector interactions, alluded to in Section 2, will not occur in this relatively simple version of QTD.

One can here define "retarded" and "advanced" tachyonic Green's function: $S_{R,A} = (iM - \gamma \cdot \partial)\Delta_{R,A}$, with:

$$\Delta_R(z) = \int \frac{d^4p}{(2\pi)^4} \frac{\theta(|\vec{p}| - M)e^{ip\cdot z}}{p^2 - M^2 - i\epsilon s(p_0)} \tag{14.29}$$

where $s(p_0) = p_0/|p_0|$; a similar representation, but with the sign of $s(p_0)$ reversed, defines $\Delta_A(z)$. Both Green's functions satisfy: $(M^2 + \partial^2)\Delta_{R,A}(z) = \delta(z_0)\hat{\delta}(\vec{z})$, so that the $S_{R,A}$ satisfy: $(iM + \gamma \cdot \partial)S_{R,A}(z) = \delta(z_0)\hat{\delta}(\vec{z})$. Both Δ_R and S_R vanish for $z_0 < 0$, and hence merit the subscript R; and conversely for Δ_A and S_A, which vanish for $z_0 > 0$. This formalism permits physical significance to be maintained for orthochronous Lorentz transformations, with a well defined sense of time and of a particle's sign of energy, while restricting consideration to Lorentz transformations that are performed outside the light cone.

From the above discussion, the initial step of Symanzik's derivation [Symanzik (1960)] of the functional reduction formula, between every S-Matrix clement and the generating functional, is valid. It is:

$$\psi_T(x) = \sqrt{Z_2}\psi_T^{\mathrm{IN}}(x) + \int d^4y\; S_R(x-y)\mathcal{D}_y\psi_T(y) \tag{14.30}$$

where $\mathcal{D}_y = iM + \gamma \cdot \partial_y$, M is the renormalized (physical) tachyon mass, $\psi_T(x)$ denotes the fully interacting (Heisenberg representation) tachyon operator, and ψ_T^{IN} is its free field counterpart, bearing its renormalized mass. Z_2 is the tachyon wave function renormalization constant. Since $\mathcal{D}_x\psi_T^{\mathrm{IN}} = 0$, the operation of $\mathcal{D}_x$ on both sides of (14.30) yields a simple identity. Once the validity of (14.30) is appreciated, all of Symanzik's functional steps follow through, with the result:

$$S/ < S[A^{\mathrm{ext}}] >= \; : \exp\left\{Z_3^{-1/2} \int A_\mu^{\mathrm{IN}}(-\partial^2)\frac{\delta}{\delta j_\mu} \right.$$
$$\left. + Z_2^{-1/2} \int \left[\bar{\psi}_T^{\mathrm{IN}}\gamma_5\vec{\mathcal{D}}\frac{\delta}{\delta\bar{\eta}} - \frac{\delta}{\delta\eta}\overleftarrow{\mathcal{D}}\gamma_5\psi_T^{\mathrm{IN}}\right]\right\} : \mathcal{Z}[\eta,\bar{\eta},j]|_{\eta,\bar{\eta},j\to 0}. \tag{14.31}$$

With (14.31), it is possible, in principle, to calculate the amplitude of any process to any perturbative order; in addition, and as in certain limiting situations in ordinary QFT, it may be possible to sum subsets of classes of Feynman graphs, with each class containing an infinite number of graphs.

14.4 Photon Emission and Reabsorption

There is one important difference between the Physics of certain "modified bremsstrahlung" processes in QED and QTD which is worth mentioning. In conventional S-Matrix descriptions of reactions initiated by a few particles, the initial and final states are understood to be asymptotically well separated, and this is expected in a world where all massive particles travel with $v < c$. But when charged tachyons and photons are involved, the situation is not as clear, because physical "overlaps" of these particles can exist for macroscopic times; and the final, physical result can be quite unexpected.

As an example, consider the amplitude for a charged tachyon of momentum T_μ to emit a photon, and to leave the scene of the reaction with momentum T'_μ. Kinematically, this process resembles Cerenkov radiation, with $T = T' + k$. Squaring and summing both sides of the relation $T' = T - k$, with $T^2 = T'^2 = M^2$, $k^2 = 0$, leads to the restrictions: $T \cdot k = \vec{T} \cdot \vec{k} - T_0 k_0 = 0$. In other words, the angle between the photon's spatial momentum $\vec{k}$, and the direction, $\hat{T}$, of the emitting tachyon, is given by $\cos\theta = T_0/|\vec{T}|$; and similarly, $\cos\theta' = T'_0/|\vec{T}'|$ defines the angle between the outgoing $\vec{k}$ and $\vec{T}'$. Note that these kinematical relations are independent of the energy of the emitted photon. Using standard Rules, as sketched in Appendix F, one can calculate the probability per unit time τ for the emission of a photon of arbitrary polarization within a band of energy ω to $\omega + d\omega$, as:

$$\frac{1}{\tau}\frac{d}{d\omega}\sum |< T', k|S|T >|^2 = \frac{\alpha M^2}{|\vec{T}|T_0} \tag{14.32}$$

where $\alpha = e^2/4\pi$. (A differential cross-section is not appropriate here, since the angular distributions are fixed by the kinematics above). With a physical upper cut-off chosen as $\omega_{\max} = T_0$, the lowest order probability/time for such emission is:

$$\frac{1}{\tau}\sum |< T', k|S|T >|^2 = \frac{\alpha M^2}{|\vec{T}|}. \tag{14.33}$$

By direct calculations, (14.33) has exactly the same probability/time for the inverse process, the absorption of a photon k by a tachyon T which results in T'. In order for this inverse process to become significant, the incident tachyon should move in a directed photon beam, at just the right kinematical angle, for an amount of time $\tau = |\vec{T}|/\alpha M^2$, using the perturbative result for the inverse process as indicative of that of a more precise calculation. Because the tachyon moves faster than c, the question to be posed is whether the outgoing photon of the emission process stays "in close proximity" to the outgoing tachyon for a duration of time of this magnitude, and can therefore be reabsorbed by the outgoing tachyon. The angles for emission and absorption for the two processes are identical, and therefore, the reabsorption of that emitted photon is a kinematic possibility. The question to be answered is: how long are the outgoing photon and tachyon "in close proximity"?

The relative velocity of emitted photon and final tachyon is the relevant quantity. With $c = 1$ (and $\hbar = 1$), the velocity vector of the emitted photon is k, while that of the final tachyon is $\vec{v}_T' = dT_0'/d\vec{T}' = \vec{T}'/T_0'$, with $T_0' = \sqrt{\vec{T}'^2 - M^2}$. The relative velocity of the two is $\vec{v}_{\text{rel}} = \hat{k} - \vec{T}'/T_0'$. Because of the spatial, kinematical relations stated above, the projection of $\vec{v}_{\text{rel}}$ alone the photon's direction vanishes:

$$\hat{k} \cdot \vec{v}_{\text{rel}} = 0. \tag{14.34}$$

Also, the projection of $\vec{v}_{\text{rel}}$ along the tachyon's direction is small:

$$\hat{T}' \cdot \vec{v}_{\text{rel}} = \hat{k} \cdot \hat{T}' - \frac{|\vec{T}'|}{T_0'} = \frac{T_0'}{|\vec{T}'|} - \frac{|\vec{T}'|}{T_0'} = -\frac{M^2}{T_0'|\vec{T}'|} \tag{14.35}$$

for a high energy tachyon, $T_0' \gg M$ (for which case $T_0 \gg M$ as well). A better indication of their relative velocity might simply be

$$\vec{v}_{\text{rel}}^2 = \left(\hat{k} - \frac{\vec{T}'}{T_0'}\right)^2 = 1 + \frac{\vec{T}'^2}{T_0'^2} - 2\frac{\hat{k} \cdot \vec{T}'}{T_0'} = +\frac{M^2}{T_0'^2} \tag{14.36}$$

again using the kinematic relation $T_0' = \hat{k} \cdot \vec{T}'$. During a time τ one would then expect the photon and tachyon to "move apart" by a distance $D \sim \tau(M/T_0')$. The significance of equations (14.34) - (14.36) is that for a high energy tachyon, with $T_0 \gg M$ and $T_0' \gg M$, the emitted photon and final tachyon separate very slowly. Using (14.36) as a measure of their velocity of relative separation, and the time $\tau \sim |\vec{T}'|/\alpha M^2$ as a qualitative measure of the time suggested by (14.33) for the probability of reabsorption to be $\sim 0(1)$, the distance of that separation grows to $D = v_{\rm rel}\tau \lesssim \frac{|\vec{T}'|}{T_0'}/\alpha M \sim 10^{-18}$ cm, for $M \gtrsim 10^7$ GeV and $T_0' \gg M$. If a value larger than 1/137 is used for α, or a larger value for M, this distance D is even shorter.

Contrast this with the distance moved by the tachyon, of velocity $|\vec{T}'|/T_0'$, during the same interval τ: $D_{T'} = v_{T'}\tau \sim \vec{T}'^2/\alpha M^2 T_0'$, which is a factor $|\vec{T}'|/M$ larger that D. The final, high energy tachyon therefore moves through a distance which is considerably larger than the separation distance between it and the emitted photons and this suggests that the photon and final tachyon remain "in close proximity", such that during this time the photon is reabsorbed by that tachyon. At least, there should be a non-zero probability of reabsorption; and in this way, over asymptotic times and distances, there should be a significant, non-zero probability that a very high energy tachyon will reabsorb every photon that it emits, and thereby retain its original high energy. If not exactly, then almost so.

Can a charged tachyon moving through an environment of Cosmic Microwave Background photons absorb those CMB photons which it meets at just the proper angle? There is no reason why this cannot happen, and it reinforces the notion that a charged tachyon moving through galactic space can achieve — and will maintain — an ever increasing energy, with a velocity just slightly larger than c.

Another argument leads to a similar conclusion. Consider the external $\vec{E}$ and $\vec{B}$ fields produced by a charged tachyon moving at constant velocity. Any external field can be represented as an integral over virtual photon modes, with weighting depending on the nature of the particular external field. If that field vanishes when viewed by an observer at the Cerenkov-like angle, it means that no photons are emitted at that angle. But emission at that angle is precisely the definition of a real photon, since then and only then can $k^2 = 0$; if the $\vec{E}$ and $\vec{B}$ fields of a constant velocity tachyon do vanish when viewed at this angle, such a classical result indicates that no real photon can be emitted by a tachyon in a Cerencov-like process. A straightforward calculation of those fields shows that, if assumed continuous, they indeed must vanish, precisely and uniquely at that angle.

14.5 Kinematic of Tachyon-Particle Reactions

The conjecture of the preceding paragraph underlies the reactions of this section, in which high energy tachyons are able to retain their energy/momenta while traveling to galactic distances, where they annihilate and/or scatter with photons and other, ordinary matter.

One does not have the freedom to make Lorentz transformations that "simplify" calculations by moving to a Lorentz frame in which the T_0 component of one tachyon vanishes, for such a transformation assigns that tachyon an infinite velocity. This does not make any physical sense; the mathematical equivalent of that step is dividing by zero.

Imagine that a charged $T - \bar{T}$ pair is produced at some space-time point, either by a Schwinger mechanism following from quantized QED vacuum fluctuations at extremely small distances [Fried (2003)], or by some other mechanism associated with the Big Bang, or subsequent, cataclysmic events. These tachyons immediately separate, each with $v > c$, moving away from the world of ordinary particles into the space between stars, and into the outer space reaches of every galaxy. But galactic magnetic fields exist, which must influence these charged tachyons in the conventional way, by bending their trajectories into partially "circular" paths, depending on the direction and magnitude of the magnetic fields encountered. (If there are a sufficient number of such tachyons, they themselves can be thought of as generating at least a portion of the galactic magnetic fields; but that subject is reserved for a separate treatment). The picture that emerges is of "swarms" of charged tachyons, moving not quite randomly in curved orbits at galactic distances; and every so often, depending on the density of such tachyons and of stray photons and bits of ordinary matter, there will occur collisions.

14.5.1 *Blocking the reaction* $\gamma \to T + \bar{T}$

We here suggest a simple, intuitive mechanism for preventing the decay of real, mass-shell photons into $T - \bar{T}$ pairs, a mechanism which may be related to the conventional Higgs–Weinberg–Salam symmetry breaking. One imagines a ψ^4 type contribution to the basic interaction Lagrangian, $\mathcal{L}'' = -(\lambda^2/2)(\bar{\psi}\gamma_5\psi)^2$, or a less singular relative (such as: $i\lambda\bar{\psi}\gamma_5\psi\chi - 1/2[\mu^2\chi^2 + (\partial\chi)^2]$) which provides a strong, short-range, binding interaction between T and $\bar{T}$, such that the excitation of a vacuum state $T\bar{T}$ pair by the Schwinger mechanism requires the addition of a sizable energy, B, to overcome this assumed $T - \bar{T}$ binding. Were a photon able to lift a $T - \bar{T}$ pair from the vacuum, it also would be required to supply B into the conservation equations:

$$k_0 = B + T + \bar{T}, \qquad \vec{k} = \vec{T} + \vec{\bar{T}}.$$

For these mass-shell particles, $k^2 = \vec{k}^2 - k_0^2 = 0$, $T^2 = \bar{T}^2 = M^2$; and if the angle between the supposedly emitted T and $\bar{T}$ is θ, it is straightforward to show that:

$$\cos^2\theta = \frac{[T_0\bar{T}_0 - M^2 + B^2/2 + B(T_0 + \bar{T}_0) + (T_0^2 + \bar{T}_0^2)/2]^2}{[M^2 + T_0^2][M^2 + \bar{T}_0^2]}. \qquad (14.37)$$

But, for example, if $B = 2M$, a little algebra shows that $\cos^2\theta > 1$, for arbitrary T_0, $\bar{T}_0$; and hence this process is forbidden.

A corollary to this argument is the observation that an extra amount of energy B is released by the annihilation reaction: $T + \bar{T} \to \gamma + \gamma$, as mentioned in connection with "loop annihilation" at the close of Section 6. If $B \sim 2M$, there would be an additional and large amount of energy released in each annihilation, in addition to the intrinsic T_0 and $\bar{T}_0$ energies of each of the annihilating tachyon pair.

14.5.2 *Scattering: $T + p = T' + p'$*

Suppose that a particle (e.g., a proton floating about in outer galactic space) is struck by an energetic tachyon. We idealize this situation to a proton at rest struck by a tachyon of energy $T_0 \gg m$, write the conservation laws in the form $T + p - p' = T'$, square and sum both sides to find an exact, linear equation (assuming $p_0' > m$):

$$p_0' = \frac{N}{D}m = \frac{(m + T_0)^2 + (M^2 + T_0^2)\cos^2\theta}{(m + T_0)^2 - (M^2 + T_0^2)\cos^2\theta}m \qquad (14.38)$$

where θ is the angle between $\vec{T}$ and $\vec{p}'$. Since $N > 0$, one expects that the largest value of p_0' will result from the smallest value of D, or the maximum possible value of $\cos^2\theta$. From energy conservation, $p_0 + T_0 - p_0' = T_0' > 0$, which when applied to the present case of $p_0 = m$, requires:

$$\cos^2\theta < \frac{T_0(m + T_0)^2}{(M^2 + T_0^2)(2m + T_0)} \equiv \cos^2\theta_{\text{max}} < 1. \qquad (14.39)$$

Substitution of the $\cos^2\theta_{\text{max}}$ of (14.39) into (14.38) then yields:

$$p_0' = m\left[\frac{1 + \frac{1}{1+2m/T_0}}{1 - \frac{1}{1+2m/T_0}}\right] \simeq T_0\left(1 - \frac{m}{T_0} + \cdots\right). \qquad (14.40)$$

An effective "billiard ball" collision has occurred, with the final proton taking on the energy of the incidental tachyon.

This example shows that an initial tachyon is able, by scattering, to transfer a considerable amount of its energy to a particle initially at rest (or moving at a modest $v < c$), and in the process create high energy cosmic rays. Radiation from the scattering of charged particles by this mechanism should, in principle, contribute X-ray and γ-ray backgrounds.

In addition, here is a mechanism which explains how occasional ultra-high-energy cosmic rays, specifically protons, can have energies exceeding the GZK limit [Greisen (1966)],[Zatsepin (1966)], [Daigne (1999)], [Piran (2004)], [Fontera (2004)]. A stray, high energy charged tachyon, approaching the Earth, strikes and transfers a large portion of its energy and momentum to a low-energy proton floating just above the atmosphere, which proton promptly initiates the observed cosmic-ray shower. The GZK limit does not apply to the tachyon, while the proton obtains its extremely high energy just before initiating that shower.

14.6 From Ehrenfest's Theorem to Loop Annihilation

It has been suggested above that charged tachyons of very high energy could reabsorb any photon emitted, thereby preserving their four-momentum across galactic times and distances; and, as they move away from and between galaxies, such tachyons could be influenced by galactic magnetic fields (of orders of magnitude from milli to micro Gauss) and caused to move in "circular" paths, at least while they are under the influence of a coherent magnetic field.

A fundamental question arises as to just how a charged tachyon would be affected by specified electromagnetic fields. There are no "classical" tachyons in our world, upon which experiments could be performed to yield Classical Tachyon Dynamics (CTD), and so the analysis must proceed in the opposite sense: from the assumed QTD Lagrangian of (14.11) to a one-particle representation of that Lagrangian, and thence to a subsequent approximation recognized as representing the "classical" action of a specified electromagnetic field on a single, charged tachyon. These last steps may be realized by a derivation of the corresponding Ehrenfest's theorem, and that is the subject of this Section.

Begin first with the free tachyon field operator of (14.7), and consider is one-particle matrix element:

$$\phi(x) = < 0|\psi_T(x)|\vec{p}, s >$$

which satisfies the same equation as does $\psi_T(x)$, and has its solution (see (G.8)):

$$\phi(x) = e^{ip\cdot x} u_s(p).$$

Because $u_s(p)$ satisfies: $(M + \gamma \cdot p)u_s(p) = 0$, upon multiplication by $i\gamma_4$ one obtains:

$$(i\gamma_4 M + \vec{\alpha} \cdot \vec{p})u_s(p) = p_0 u_s(p)$$

which permits the identification of $H_T^{(0)} = iM\gamma_4 + \vec{\alpha} \cdot \vec{p}$ as the one-free-tachyon Hamiltonian operator, with real eigenvalue p_0. In QED, the same analysis provides the one-free-electron Hamiltonian, $H^{(0)} = m\gamma_4 + \vec{\alpha} \cdot \vec{p}$.

The gauge invariant extension of this one-tachyon Hamiltonian to include the effects of an external electromagnetic field is immediate:

$$H_T \to iM\gamma_4 + \vec{\alpha} \cdot (\vec{p} - e\vec{A}) + eA_0,$$

just as one does in QED.

The time dependence of the one-tachyon wave function, assumed properly normalized according as: $\int d^3x \phi^\dagger \gamma_5 \phi = 1$ is given by the relevant Schrodinger equation $i\hbar\, \partial\phi/\partial t = H_T\phi$, and its adjoint $-i\hbar\, \partial\phi^\dagger/\partial t = \phi^\dagger H_T^\dagger$, while the expectation value of any rachyonic operator Q is defined by:

$$< Q > = \int d^3x\, \phi^\dagger \gamma_5 Q \phi.$$

As usual in this "Schrodinger representation", the full time dependence is carried by the wave functions, and the operators Q are understood to be time independent. But the c-number $A_\mu^{\text{ext}}(x)$ are allowed to depend upon time, since unitary transformations on c-numbers cannot effect their time dependence. Hence, the total time rate of change of $< Q >$ is given by:

$$\frac{d < Q >}{dt} = \int d^3x\, \phi^\dagger \gamma_5 \left(\frac{\partial Q}{\partial t} + \frac{1}{i\hbar}[Q, H_T] \right) \phi$$

where, since $\{\gamma_5, \gamma_4\} = 0 = [\gamma_5, \vec{\alpha}]$, $H_T = \gamma_5 H_T^\dagger \gamma_5$. Thus, if $Q \to \vec{r}$,

$$\frac{d<\vec{r}>}{dt} = \int d^3x\ \phi^\dagger \gamma_5 \vec{\alpha} \phi$$

showing, as in QED, that $\vec{\alpha}$ is the operator whose expectation value corresponds to velocity.

The average momentum of a charged tachyon in an external electromagnetic field is calculated from the gauge invariant, "canonical" contribution:

$$<\vec{p}>= \int d^3x\ \phi^\dagger \gamma_5 (\vec{p} - e\vec{A})\phi$$

and one easily finds that:

$$\begin{aligned}\frac{d<\vec{p}>}{dt} &= -e \int d^3x\ \phi^\dagger \gamma_5 \frac{\partial \vec{A}}{\partial t}\phi + \frac{1}{i\hbar}\int d^3x\ \phi^\dagger \gamma_5 [(\vec{p} - e\vec{A}), H_T]\ \phi \\ &= e \int d^3x\ \phi^\dagger \gamma_5 \left[-\vec{\nabla} A_0 - \frac{\partial \vec{A}}{\partial t} + \vec{\alpha} \times \vec{B} \right] \phi \\ &= e \int d^3x\ \phi^\dagger \gamma_5 \left[\vec{E} + \vec{\alpha} \times \vec{B}\right] \phi. \end{aligned} \tag{14.41}$$

Exactly the same relation, but without the γ_5 factor, is obtained in QED.

If we now suppose that $\vec{E}$ and $\vec{B}$ are spatially constant over the dimensions of the tachyon — or, more precisely, over the wave packet that describe the tachyon, as it moves with $v > c$ — then (14.41) can be rewritten as:

$$\frac{d<\vec{p}>}{dt} = e\left[\vec{E} + <\vec{v}> \times \vec{B}\right]$$

which is the conventional relation for a charged particle moving in specified $\vec{E}$ and $\vec{B}$ fields, with $\vec{p}$ and $\vec{v}$ now replacing $<\vec{p}>$ and $<\vec{v}>$. If we wish to write the corresponding four vector as:

$$p_\mu = M \frac{dx_\mu}{d\tau} = M u_\mu$$

where $\sum_\mu u_\mu^2 = +1$ (rather than the -1 of the Minkovski metric) and:

$$\frac{dp_0}{d\vec{p}} = \frac{\vec{p}}{p_0} = \vec{v}$$

the covariant equation for the motion of a charged particle is again valid, as in QED:

$$\frac{d^2x_\mu}{d\tau^2} = \frac{e}{M}F_{\mu\nu}(x)\frac{dx_\nu}{d\tau}$$

except that the tachyon mass M replaces the electron's m, and $dt/d\tau = \gamma = 1/\sqrt{v^2 - 1}$.

With this demonstration that the motion of a charged tachyon follows essentially the same classical equations as those of other charged particles, it is now possible to understand how such charged tachyons can contribute to Dark Matter. Consider the production of a burst of charged Ts, and $\bar{T}$s, by some catastrophic process. They separate, presumably moving in roughly opposite directions; and, concentrating on the Ts, assume they possess a distribution of velocities moving roughly in the same direction. The lead T of this group, moving with the fastest velocity, generates a magnetic field which it outruns and cannot feel; but the next T behind it will be affected by that field, and deflected so that it eventually lines up behind the first T. And the same process happens all way down the line, resulting in a "line", or effective current of such charged Ts.

Imagine that this current line is moving through galactic space — the space surrounding galaxies — where there exist magnetic fields of magnitude $B \sim 10^{-6}$ Gauss, which are coherent across distances $R \sim 10^4$ light years. One naturally expects such Ts to fall into orbits defined (reinstating the c dependence) by:

$$Mv^2/R = \gamma^{-1}evB/c \tag{14.42}$$

where $\gamma^{-1} = [(v/c)^2 - 1]^{1/2}$, and $v > c$. For high energy Ts, $v \gtrsim c$, and so let us choose $(v/c)^2 = 1 + 10^{-6}$, $\gamma = 10^3$, and insert the observed B and R values to obtain from (14.42) an estimate of what we might expect for M. Amusingly, the result is $M \simeq 10^7$ GeV, which is just the order of magnitude found in the original calculation of [Fried (2003)].

The point of this argument is that if those B fields arise from currents in the visible galactic matter, then the charged Ts trapped in orbits by those Bs are "rigidly" connected to the matter that we can see; but the Ts themselves are invisible, they are "dark". (Note that were any coherent cyclotron radiation possible by such loops, its typical frequency would be far too small to detect, $\omega = v/R \simeq 10^{-14}s^{-1}$). If there are enough such charged Ts, they could contribute to the dark, or missing matter of current observation. This argument can be easily generalized to include Ts of all energies.

Further, if there exist two or more such loops, of roughly the same size, parallel to each other and separated horizontally by a distance on the order of their radius, then their own magnetic fields would combine with that of the B which hold them to create a "magnetic bottle" effect, trapping any available number of ordinary charged particles, which would then also contribute to dark matter. Two such loops, each with their "mega- macroscopic" magnetic moments pointing (eventually) along the magnetic field $\vec{B}$ would form a stable system: their magnetic moments attract but their like electric charges repel each other, leading to an equilibrium separation distance of the order of their loop radius.

Finally, if one imagines that one loop consists of $\bar{T}$ circulating in the opposite direction to that of the T loop, their loop magnetic moments point in the same direction along the line separating them, and the T and $\bar{T}$ loops would be attracted towards each other — both their magnetic moments and unlike charges attract — eventually leading to a cataclysmic "loop annihilation", which might even be visible astronomically. If the T and $\bar{T}$ loops encircle a large gas cloud, as they well might, then the "perfect fusion", loop-annihilation could emit sufficient radiation and newly formed high velocity matter to irradiate and compress the gas, setting off internal shock waves, and perhaps even a subsequent nuclear reaction. The corresponding γ-ray production, with X-ray and optical tails, and a natural afterglow, could possibly match the well known, observed GRB data [Greisen (1966)], [Zatsepin (1966)], [Daigne (1999)], [Piran (2004)], [Fontera (2004)].

14.7 NASA's Fermi Bubbles

The "Loop Annihilation" of the previous section suggests a possible mechanism for understanding the recent observations [Finkbeiner, et al. (2010)] of two large, magenta-colored "bubbles", one above and one below the center of our galactic plane, emitting radiation from the optical to the multi-GeV. The radii of these bubbles each appear to be on the order of a quarter of the galactic radius; and during the course of observations the boundaries of the bubbles appear well-defined and unchanging. Obvious questions immediately arise: Where does such a huge amount of radiant energy come from? Why are the bubble radii so large? And why are their boundaries apparently fixed?

A tentative, possible explanation would appear to follow from the "Loop Annihilation" mentioned at the close of the previous section. For clarity and simplicity, assume that a T- and a $\bar{T}$- loop with effective magnetic moments perpendicular to, and surrounding a good portion of the galactic plane, find themselves about to undergo annihilation. A two-dimensional, "fusion explosion" will take place, with enormous amounts of energy liberated, including the energy that was originally needed to unbind the tachyon pairs and tear them out of the quantum vacuum. Were the two loops not perfectly aligned, the annihilation would not be instantaneous, but the deposition of a huge amount of energy would still occur at space-time points corresponding to the intersection of the two loops; and in this way the first of the three questions stated above may be answered.

For simplicity, imagine that the two loops are of the same diameter, and that the annihilation happens instantaneously. At each point on the rim of this fiery circle where a fusion reaction is taking place, each with the emission of a pair of highly energetic gammas, energy is being radiated in all directions, out into space and, in particular, towards the center of the galactic plane. This is happening over vast distances, on the order of kiloparsecs, and the gas clouds, mostly protons, caught inside that circle will begin to feel the pressure of that radiation, with the result that they will tend to move into regions of lower pressure, above and below the galactic plane.

But the annihilation radiation contains multi-GeV gammas; and this radiation, directed inwards from around the galactic plane, and striking the protons near and slightly above and below that plane, will have the effect of splitting apart a large number of protons into their quark/gluon constituents. The quarks and antiquarks so produced are pulled by myriad gluon flux tubes, trying to reform into nucleons and mesons, and in the process are radiating photons over a wide spectrum of energies, including energetic gammas, some of which leave the region and others which serve to split apart nucleons and mesons which have just formed. In this way, such bubbles appear to be more or less stationary.

What is it that supports such a conclusion? It is the observation that the boundaries of each bubble appear well-defined; radiation, of optical frequencies and gammas of GeV energies, is measured coming from within the observed boundaries, which appear to be fairly stable. What dynamical reason is there that such stability should exist? It is simply that a quark trying to leave the bubble will be pulled back by the gluonic flux tubes connecting it to other quarks and antiquarks inside that bubble; it is as if the bubble possesses an enormous surface tension, such that none of its fundamental constituents can escape.

However, one obvious objection to this picture is that quark-gluon plasmas are highly dense constructs, taking place over less than nucleon sizes. Even if there are a huge number of $T - \bar{T}$ pairs annihilating, producing huge numbers of multi-GeV gammas, the visible size of the bubbles is far too large to be formed by the partial remnant of isolated gas clouds which happened to find themselves inside the circle of annihilating tachyons. Conceptually, one needs a source of matter far more dense and large and available than an isolated gas cloud.

And then one realizes that there is a black hole at the center of our galaxy, continuously drawing matter into itself; and here is a possible source of matter, under continuous radiation from the first, less dense pair of bubbles. In effect there are two sources, one above and one below the galactic plane, whose ordinary matter is being subjected to multi-GeV bombardment, from both the original tachyon annihilation, and the continuing recombinations of multiple quark-antiquark pairs, whose GeV radiation then serves to split apart other matter in the bubbles as well as matter being drawn into the black hole region. And now, with a sufficient density of hadronic matter being drawn into the bubbles, originally on its way into the black hole, one can imagine the Fermi bubbles as Condensing Quark-Gluon plasmas of enormous size.

As with any Model of physical phenomena, one requires serious estimates of relevant parameters, so as to know if it should be take seriously. This the author must leave to astrophysicists far more learned than himself. Time will tell if this or any other model can be accepted as 'probably true'; but regardless of that conclusion, it is wise and useful to remember that these Astrophysical Speculations began with a theory of the very small, and ended with a possible description on a galactic scale. That is the power and the reach of Quantum Field Theory.

14.8 Summary

In the belief that any model which can simultaneously relate astrophysical observation results involving vacuum energy, dark matter, and the origin of ultra-high-energy cosmic rays and gamma bursts is worth considering, we have here put forth the idea of charged tachyons, as a possible, underlying mechanism. Our charged tachyons are fermionic, rather than scalar, to minimize the probability of Cerenkov-like radiation by the tachyons over galactic distances.

If the tachyons do not couple directly to pions, but only through higher order photonic interactions, then the Greisen–Zatsepkin–Kuzmin cut-off of high-energy cosmic rays is effectively removed, compared to the case of energetic protons; and only after a collision of a very high energy tachyon with a slow proton — resulting in that proton's absorption of a sizable fraction of that tachyon's energy — does the standard GZK cut-off come into play. But if that newly accelerated proton lies in our upper atmosphere, and fairly close to Earth, it can effectively violate the GZK limit.

Precisely how much of a contribution to Dark Matter could be made by such tachyons depends on the knowledge of several points not yet clear: the rate of charged tachyon pair production by the Schwinger mechanism, and possible sources of electromagnetic fields to induce such production; the average number of tachyons per galactic loop, and the average number of such loops per galaxy. Estimates of some of these questions are now underway; but at the moment, all that one can reasonably say is that such charged tachyons provide a mechanism which could contribute to the understanding of these current, astrophysical problems. One hopes that the QTD analysis sketched above will generate further efforts in this direction.

In addition to the topics mentioned above, it would be of considerable interest to understand, in detail, the following items.

(1) Possible radiation signatures from the scattering of tachyons on ordinary particles, at galactic distances, especially if the tachyon charge is different from the charge of the particle it strikes.
(2) Tachyonic emission and absorption of CMB photons.
(3) The use of these charged tachyons as a possible condensation-mechanism for the Weinberg–Salam vacuum.
(4) The incorporation of such tachyons into a General Relativistic framework.

One final remark (in response to constructive criticism of Professor J. Avan). One may feel uncomfortable without a proper definition of the tachyon vacuum state, in terms of conventional particle theory. Let us ask the familiar question what is the mass of a quark? The well-known answer is that it is physically impossible to measure directly its mass, its intrinsic energy; and that one must be content with estimates that can be inferred from measurements which display results that depend on the type of measurement. If quantum mechanics teaches us anything, it is that no statements should be made — or can be believed — about quantities that cannot be measured. But we infer that a quark must have an intrinsic mass, and we assume that a quark vacuum state exists.

In an analogous way, one cannot directly measure any properties of our assumed tachyons, including their absence; but they can be used to explain, or to contribute to the understanding of certain, puzzling, astrophysical observations. Does our lack of ability for direct measurements mean that there exists no tachyon vacuum? Not at all; it merely reflects the fact that it is impossible for us to measure directly the properties of an isolated, individual tachyon, just as for the quark situation, but for different kinematical and dynamical reasons. Again, all we can do is to infer the properties of tachyons by matching their predictions against astrophysical observations.

Acknowledgment

Thanks are due to many people for their aid and interest in formulating the contents of this book.

The author is indebted to his immediate colleagues and co-authors — Y. Gabellini, T. Grandou, and Y-M Sheu, Institute Non-Lineaire de Nice — for the collaborative research discussed in the papers forming the material of Parts 2, 3 and 4. Thanks are also due to M. Gattobigio, INLN, for his informative conversations relevant to the Nuclear Physics aspects of this work.

For a variety of most useful conversations, the author warmly thanks Professors I. Dell'Antonio, S. Koushiappas and Doctor W. Becker of Brown University; Professor J. Avan of l'Universite de Cergy-Pontoise and Doctor F. Daigne of l'Institut d'Astrophysique de Paris.

Warm thanks are due to Sabina Griffin for her cheerful and superb typing of the entire manuscript. It is also a pleasure to thank Marc Rostollan, of the American University of Paris, for his kind assistance in arranging sites for our collaborative research, when in Paris.

This publication was made possible through the support of a Grant from the John Templeton Foundation. The opinions expressed in this publication are those of the author and do not necessarily reflect the views of the John Templeton Foundation.

Appendix A

Equivalence Example of the DP Model

To demonstrate the accuracy of the DP Model in a quite different setting, consider the calculation of the simplest fermion self energy graph, where — to make connection with the boson nature of the $L[A]$ calculations — we simplify by passing to a bosonized version of relevant fermion propagator at the same stage of each calculation. This example shows that the log divergences of both calculations are exactly the same.

The g^2 order contribution to the fermion propagator is given, in functional notation, by:

$$-\frac{i}{2}\int \frac{\delta}{\delta A_\mu} D_c^{\mu\nu} \frac{\delta}{\delta A_\nu} G_c(x,y|A)\big|_{A\to 0}$$
$$= (-i)(ig)^2 \int D_c(u-w) G_c(x,u|A)\gamma_\mu G_c(u,w|A)\gamma_\mu G_c(w,y|A)\big|_{A\to 0} \tag{A.1}$$

where we again adopt the simplest Feynman gauge $D_{c,\mu\nu} = \delta_{\mu\nu} D_c$.

For the Feynman graph calculation, we replace both of the external G_c by their free field limit S_c, and, to have a clear correspondence with the DP calculation, imagine that the central G_c is replaced by a bosonized version, $< u|[m^2 + (\partial - igA)^2]^{-1}|w >$, whose free particle limit is simply $< u|[m^2 + \partial^2]^{-1}|w >$. Taking Fourier transforms, one finds the corresponding contribution to (A.2):

$$\frac{ig^2}{(2\pi)^4} \tilde{S}_c(p)\gamma_\mu \int d^4k \frac{1}{k^2 - i\epsilon} \frac{1}{(p-k)^2 + m^2} \gamma_\mu \tilde{S}_c(p)$$

or, with $\tilde{S}_c(p) = \frac{1}{m+ip}$, $\sum_\mu \gamma_\mu\gamma_\mu = 4$:

$$-\frac{ig^2}{4\pi^4}\left(\frac{1}{p^2}\right) intd^4k \frac{1}{k^2 - i\epsilon} \frac{1}{(p-k)^2 - i\epsilon} \tag{A.2}$$

where, for simple comparison with the DP Model, we choose $m = 0$.

Using the simplest regularization possible:

$$\frac{1}{k^2} \to \frac{1}{k^2} - \frac{1}{k^2 + \Lambda^2} = \int_0^{\Lambda^2} \frac{dl}{k^2 + l^2} \tag{A.3}$$

and the Feynman combinatoric $\frac{1}{a^2 b} = 2\int_0^1 dx(1-x)[a + x(b-a)]^{-3}$, the integral of (A.3) is easily shown to be: $i\pi^2 \int_0^1 dx \ln\left(\frac{\Lambda^2}{xp^2}\right)$.

The result of this computation is then:

$$\frac{g^2}{4\pi^2}\frac{1}{p^2}\int_0^1 dx \ln\left(\frac{\Lambda^2}{xp^2}\right). \tag{A.4}$$

We now calculate the equivalent quantity using the DP Model of the fermion propagator, in which all spinorial terms are suppressed, and $m = 0$:

$$\begin{aligned} &i\int_0^\infty dse^{-ism^2} \cdot N \cdot e^{-1/2Tr\ln(2h)} \int d[u] e^{\frac{i}{4}\int u(2h)^{-1}} \delta^{(4)}(u(s) + z) \\ &\times \frac{ig^2}{2}\frac{i}{4\pi^2}\int_0^s ds_1 \int_0^s ds_2 u'(s_1)\cdot u'(s_2)\left[(u(s_1) - u(s_2))^2 + i\epsilon\right]^{-1}. \end{aligned} \tag{A.5}$$

Using the notation and operations as defined in the above sections, the DP Model replaces the second line of (A.5) by:

$$\frac{g^2}{4\pi^2}\int_0^s ds_1 \int_0^{s_1} ds' \frac{\partial}{\partial s'}\frac{1}{s' + i\epsilon\bar{\epsilon}} = \frac{g^2}{4\pi^2}\int_0^s ds_1 \left[\frac{1}{s_1 + i\epsilon\bar{\epsilon}} - \frac{1}{i\epsilon\bar{\epsilon}}\right]. \tag{A.6}$$

We again argue that $\int_0^s ds_1 \bar{\epsilon} = 0$, and replace the integral of (A.6) by:

$$\int_{i\epsilon\bar{\epsilon}(0)}^s \frac{ds_1}{s_1} = \ln\left(\frac{s}{i\epsilon\bar{\epsilon}(0)}\right). \tag{A.7}$$

The first line of (A.5) is then immediately given by: $i\int_0^\infty ds \int \frac{d^4p}{(2\pi)^4} e^{ip\dot{z}} e^{-isp^2}$, and in combination with (A.7), its Fourier transform then yields:

$$\frac{g^2}{4\pi^2} i \int_0^\infty dse^{-isp^2} \ln\left(\frac{s}{i\epsilon\bar{\epsilon}(0)}\right). \tag{A.8}$$

Rotating the s integration contour so that $s \to -r\tau$, and with $\epsilon = \Lambda^{-2}$, $\bar{\epsilon}(0) = -1$, $x = \tau p^2$, this becomes:

$$\frac{g^2}{4\pi^2}\frac{1}{p^2}\int_0^\infty dse^{-x} \ln\left(\frac{\Lambda^2 x}{p^2}\right) \tag{A.9}$$

and one sees that the log divergent terms of (A.9) and (A.4) are exactly the same.

Appendix B

Intuitive Justification of the DP Model

One can attempt a heuristic justification of the DP Model in the following way. Because the Fradkin $v(s')$, $0 < s' < s$, refers to the four velocity of a virtual particle (e.g., the fermion moving in a closed loop of the radiative correction corresponding to the simplest vacuum bubble), we demand that $v(s')$, and therefore $u'(s')$, be a continuous function of its proper time, s'. This is a physical restriction on the class of functions $u(s')$ allowed; whether the particle is real or virtual, its four velocity should be and will be assumed to be continuous.

From its definition, $u(s') = \int_0^{s'} ds'' v(s'')$, $u(s')$ itself must be a continuous function of s'; and hence we have restricted consideration to the class of functions $u(s')$ which are continuous and have a continuous first derivative. But no statement can be made about higher derivatives, which must be expected to be discontinuous. One way to describe such functions is to imagine that they begin life as continuous, with continuous second and higher derivatives proportional to finite constant p_n. Then as the functional integration proceeds, imagine that certain of these parameters p_n are changed, in a random way, to have values which approach $\pm\infty$; and this corresponds to the introduction of discontinuous higher order derivatives. What will then be the effect of such fluctuations on the relevant functions:

$$-\frac{g_0^2}{2(2\pi)^4} \int_0^s ds_1 \int_0^s ds_2 u'(s_1) \cdot u'(s_2) \left[(u(s_1) - u(s_2))^2 + i\epsilon\right]^{-1}$$

of our functional integrand?

Consider first the denominator $\left[(u(s_1) - u(s_2))^2 + i\epsilon\right]^{-1}$. Before the p_n fluctuations are allowed to destroy the continuity of the second and higher derivatives, the main contribution of this denominator will, clearly, come from the expansion we have used, replacing $(u(s_1) - u(s_2))$ by $u'(s_1)(s_1 - s_2)$, and neglecting terms such as $(s_1-s_2)^2 u''(s_1)/2$. If the p_n are then allowed to fluctuate such that any higher derivatives approach $\pm\infty$, the denominator will then become infinite, and so conveniently removes itself from the FI.

Now consider the $u'(s_1)\cdot u'(s_2)$ numerator term, which we have replaced by $u'^2(s_1)$. Again imagine that all derivatives begin life as continuous, and consider the first correction to this approximation, $(s_2 - s_1)u'(s_1)\cdot u''(s_1)$. If $u''(s_1)$ is finite, in the neighborhood $s_2 \simeq s_1$, where the denominator is about to vanish, then this term generates a relatively small correction to our approximation; but if and when $u''(s_1)$ becomes infinite, there will be no contribution to the FI because the denominator fluctuations will involve $u''(s_1)^2$, which is more divergent than the numerator $u'(s_1)\cdot u''(s_1)$; and hence, both numerator and denominator contributions from such a discontinuous second derivative are removed from the FI.

This argument can be repeated for every higher order derivative, and for the sum of all such higher order derivatives; and the result is the DP Model defined in the test. This argument is heuristic, a physicist's argument; it is intuitive, rather than mathematically rigorous. To justify this intuition, we point to the example of Appendix A, wherein the log divergence calculated from the DP Model yields exactly the same result as the obtained from the corresponding Feynman graph integral. In this very real sense, a divergent Feynman graph in momentum space, calculated from continuous, if overlapping, momentum space integrands, may be thought of as equivalent to the "continuous" elements of a functional integrand in the DP Model.

Appendix C

Connected Cluster-Expansion Functionals

There are two distinct reasons why the connected Q_n, $n > 1$, may be expected to generate a contribution to $T(x)$ which is considerably less than that of $Q-1$. The first is because of the reduced probability of finding overlaps of three or more coordinate systems whose origins must lie within a distance (in Euclidean 4 dimensions) of $\xi\sqrt{\epsilon}$ from each other, one of those origins being that of the x' of $\frac{\delta^2 \bar{L}}{\delta A_\mu(x)\delta A_\nu(y)}$. The probability of such overlaps can be thought of as proportional to the overlapping volume divided by the total volumes of all three or more spheres each of radius $\xi\sqrt{\epsilon}$, and that ration is a small number, becoming smaller as n increases. In mathematical terms, a non-zero intersection of the supports of three or more independent distributions such as $D_c D_d \ldots D_c$ is much less than the corresponding quantity for the case of two such independent distributions.

The second reason is more closely tied to the computations of Section 6.4 and has as its origin the nature of the "connectedness" requirement, which can be stated in the following way. Connected linkages require that there shall be at least one linkage between the connected parties, e.g., for Q_2:

$$\bar{L}(e^{\overleftrightarrow{\mathfrak{D}}_A} - 1)\bar{L}\Big|_{A\to 0} \tag{C.1}$$

with the result that the linkages between the factors of $\exp\left[-ig_0\int_0^t dt' v'(t')\cdot A(x''-v(t'))\right]$ from one $\bar{L}$, and the factor $\exp\left[-ig_0\int_0^r dr' w'(r')\cdot A(x'''-w(r'))\right]$ from the other $\bar{L}$, will appear in the form:

$$\exp\left[ig_0^2\int_0^t dt'\int_0^r dr'w'(r')\cdot v'(t')D_c(x''-x'''+w(r')-v(t'))\right]-1$$

which can be rewritten as:

$$\int_0^1 d\lambda\frac{\partial}{\partial\lambda}\exp\left[i\lambda g_0^2\int\int w'\cdot v'D_c\right]$$

or as:

$$\left[ig_0^2\int\int w'\cdot v'D_c\right]\int_0^1 d\lambda\exp\left[i\lambda g_0^2\int\int w'\cdot v'D_c\right]. \tag{C.2}$$

Assume that the overlaps have taken place and consider the multiplicative term of (C.2). It does not involve $y_{1,2}$ dependence to a power, but rather the logarithm of that dependence, which under the same variable changes as used in Section (5.4), will convert to x and $u_{1,2}$ inside logs. From our over simplified model of Section (6.4), one sees that the significant x value of the final Z_3^{-1} integral is just barely larger than 1, and hence the $\ln(x)$ terms of this log dependence will not contribute significantly. The remaining $u_{1,2}$ factors must be evaluated within the $\int du_{1,2}$ of each $\bar{L}$. For $1 < u_{1,2} < x$, we get another $\ln(x)$, but for $0 < u_{1,2} < 1$, there will be non zero contribution, integrable quantities of $O(1)$, relative to the results of the $\int du_{1,2}$ integrals without such terms.

However, these log terms multiply an integral which is essentially an average over values of $0 < \lambda < 1$, over the forms $(q)^{\lambda p}$, where q takes on the different values which can be read off from the procedure, e.g., $(y_1)^{\lambda p}\left(\frac{y_1-i}{y_1-y_2-i}\right)^{\lambda p}$. Since p has been here replaced by λp, for small λ the contributions essentially disappear; and only when $\lambda\sim 1$ is there a significant value to the integral. This represents an effective decrease of the effectiveness of the coupling between the two $\bar{L}$s; and when multiplied by the small log terms discussed above, in addition to the small overlap factors, it seems clear that the connected terms cannot significantly add to $T(x)$, and therefore cannot significantly change the result of the Q_1 computation. Without a detailed and rigorous analysis, one cannot be absolutely sure; but this is one's (intuitive) belief.

Appendix D

Fradkin's Representations for non-Abelian $G_c[A]$ and $L[A]$

The exact functional representations of these two functionals of $A(x)$ are perhaps the most useful tools in all of QFT, for they allow that A-dependence of these functionals to be extracted from inside ordered exponentials; and because they, themselves, are Gaussian in their dependence upon $A(x)$, they permit the functional operations of the Schwinger/Symanzik generating functional (Gaussian functional integration, or functional linkage operation) to be performed exactly. This corresponds to an explicit sum over all Feynman graphs relevant to the process under consideration, with the results expressed in terms of functional integrals over the Fradkin variables; and in the present QCD case, because of EL, those non-perturbative results can be extracted and related to physical measurements.

The causal quark Green's function (which is essentially the most customary Feynman one) can be written as [Fradkin (1966)], [Fried (1990)]

$$G_c[A] = [m + i\gamma \cdot \Pi][m + (\gamma \cdot \Pi)^2]^{-1} = [m + i\gamma \cdot \Pi] \cdot i \int_0^\infty ds \, e^{-ism^2} e^{is(\gamma \cdot \Pi)^2}, \tag{D.1}$$

where $\Pi = i[\partial_\mu - igA_\mu^a \tau^a]$ and $(\gamma \cdot \Pi)^2 = \Pi^2 + ig\sigma_{\mu\nu}\mathbf{F}_{\mu\nu}^a \tau^a$ with $\sigma_{\mu\nu} = \frac{1}{4}[\gamma_\mu, \gamma_\nu]$. Following Fradkin's method [Fradkin (1966)], [Fried (1990)] and replacing Π_μ with $i\frac{\delta}{\delta v_\mu}$, one obtains

$$\begin{aligned}
&\mathbf{G}_c(x,y|A) \\
&= i\int_0^\infty ds\, e^{-ism^2}\cdot e^{i\int_0^s ds' \frac{\delta^2}{\delta v_\mu^2(s')}}\cdot\left[m-\gamma_\mu\frac{\delta}{\delta v_\mu(s)}\right]\delta(x-y \\
&+\int_0^s ds'\, v(s'))\times\left(\exp\left\{-ig\int_0^s ds'\left[v_\mu(s')A_\mu^a\left(y-\int_0^{s'} v\right)\tau^a\right.\right.\right. \\
&\left.\left.\left.+\, i\,\sigma_{\mu\nu}\mathbf{F}_{\mu\nu}^a\left(y-\int_0^{s'} v\right)\tau^a\right]\right\}\right)_+\Bigg|_{v_\mu\to 0}.
\end{aligned} \tag{D.2}$$

Then, one can insert a functional 'resolution of unity' of form

$$1=\int d[u]\delta[u(s')-\int_0^{s'} ds''\ v(s'')], \tag{D.3}$$

and replace the delta-functional $\delta[u(s')-\int_0^{s'} ds''\ v(s'')]$ with a functional integral over Ω and then the Green's function becomes [Sheu (2008)]

$$\begin{aligned}
&\mathbf{G}_c(x,y|A) \\
&= i\int_0^\infty ds\ e^{-ism^2}e^{-\frac{1}{2}\mathbf{Tr}\ln(2h)}\int d[u]e^{\frac{i}{4}\int_0^s ds'[u'(s')]^2}\delta^{(4)}(x-y+u(s)) \\
&\times[m+ig\gamma_\mu A_\mu^a(y-u(s))\tau^a] \\
&\left(e^{-ig\int_0^s ds' u'_\mu(s')A_\mu^a(y-u(s'))\tau^a+g\int_0^s ds'\sigma_{\mu\nu}\mathbf{F}_{\mu\nu}^a(y-u(s'))\tau^a}\right)_+,
\end{aligned} \tag{D.4}$$

where $h(s_1,s_2)=\int_0^s ds'\Theta(s_1-s')\Theta(s_2-s')$. To remove the A-dependence out of the linear (mass) term, one can replace $igA_\mu^a(y-u(s))\tau^a$ with $-\frac{\delta}{\delta u'_\mu(s)}$ operating on the ordered exponential so that

$$\begin{aligned}
&\mathbf{G}_c(x,y|A) \\
&= i\int_0^\infty ds\ e^{-ism^2}e^{-\frac{1}{2}\mathbf{Tr}\ln(2h)}\int d[u]e^{\frac{i}{4}\int_0^s ds'[u'(s')]^2}\delta^{(4)}(x-y+u(s)) \\
&\times\ [m-\gamma_\mu\frac{\delta}{\delta u'_\mu(s)}] \\
&\left(e^{-ig\int_0^s ds' u'_\mu(s')A_\mu^a(y-u(s'))\tau^a+g\int_0^s ds'\sigma_{\mu\nu}\mathbf{F}_{\mu\nu}^a(y-u(s'))\tau^a}\right)_+.
\end{aligned} \tag{D.5}$$

To extract the A-dependence out of the ordered exponential, one may use the following identities,

$$1 = \int d[\alpha]\delta[\alpha^a(s') + gu'_\mu(s')A^a_\mu(y - u(s'))]$$
$$1 = \int d[\Xi]\delta[\Xi^a_{\mu\nu}(s') - g\mathbf{F}^a_{\mu\nu}(y - u(s'))], \tag{D.6}$$

and the ordered exponential becomes

$$\begin{aligned}&\left(e^{-ig\int_0^s ds' u'_\mu(s')A^a_\mu(y-u(s'))\tau^a + g\int_0^s ds'\sigma_{\mu\nu}\mathbf{F}^a_{\mu\nu}(y-u(s'))\tau^a}\right)_+\\ &= \mathcal{N}_\Omega\mathcal{N}_\Phi \int d[\alpha]\int d[\Xi]\int d[\Omega]\int d[\Phi]\left(e^{i\int_0^s ds'[\alpha^a(s') - i\sigma_{\mu\nu}\Xi^a_{\mu\nu}(s')]\tau^a}\right)_+\\ &\times e^{-i\int ds'\Omega^a(s')\alpha^a(s') - i\int ds'\Phi^a_{\mu\nu}(s')\Xi^a_{\mu\nu}(s')}\\ &\times e^{-ig\int ds' u'_\mu(s')\Omega^a(s')A^a_\mu(y-u(s')) + ig\int ds'\Phi^a_{\mu\nu}(s')\mathbf{F}^a_{\mu\nu}(y-u)s'))},\end{aligned} \tag{D.7}$$

where $\mathcal{N}_\Omega$ and $\mathcal{N}_\Phi$ are constants that normalize the functional representations of the delta-functionals. All A-dependence is removed from the ordered exponential and the resulting form of the Green's function is exact (it entails no approximation). Alternatively, extracting the A-dependence out of the ordered exponential can also be achieved by using the functional translation operator, and one writes

$$\begin{aligned}\left(e^{+g\int_0^s ds'[\sigma_{\mu\nu}(y-u(s'))\tau^a]}\right)_+ &= e^{g\int_0^s ds'\mathbf{F}^a_{\mu\nu}(y-u(s'))\frac{\delta}{\delta\Xi^a_{\mu\nu}(s')}}\\ &\cdot\left(e^{\int_0^s ds'[\sigma_{\mu\nu}\Xi^a_{\mu\nu}(s')\tau^a]}\right)_+\Big|_{\Xi\to 0}.\end{aligned}$$

For the closed-fermion-loop functional $\mathbf{L}[A]$, one can write [Fried (1990)]

$$\mathbf{L}[A] = -\frac{1}{2}\int_0^\infty \frac{ds}{s}e^{-ism^2}\left\{\mathbf{Tr}\left[e^{-is(\gamma\cdot\Pi)^2}\right] - \{g=0\}\right\}, \tag{D.8}$$

where the trace $\mathbf{Tr}$ sums over all degrees of freedom, space-time coordinates, spin and color. The Fradkin representation proceeds along the same steps as in the case of $\mathbf{G}_c[A]$, and the closed-fermion-loop functional reads

$$\begin{aligned}\mathbf{L}[A] &= -\frac{1}{2}\int_0^\infty \frac{ds}{s}e^{-ism^2}e^{-\frac{1}{2}\mathbf{Tr}\ln(2h)}\\ &\times\int d[v]\delta^{(4)}(v(s))e^{\frac{i}{4}\int_0^s ds'[v'(s')]^2}\\ &\times\int d^4x\,\mathbf{tr}\left(e^{-ig\int_0^s ds' v'_\mu(s')A^a_\mu(x-v(s'))\tau^a + g\int_0^s ds'\sigma_{\mu\nu}\mathbf{F}^s_{\mu\nu}(x-v(s'))\tau^a}\right)_+\\ &-\{g=0\}\end{aligned} \tag{D.9}$$

where the trace **tr** sums over color and spinor indices. Also, Fradkin's variables have been denoted by $v(s')$, instead of $u(s')$, in order to distinguish them from those appearing in the Green's function $\mathbf{G}_c[A]$. One finds

$$
\begin{aligned}
\mathbf{L}[A] = & -\frac{1}{2}\int_0^\infty \frac{ds}{s} e^{-ism^2} e^{-\frac{1}{2}\mathbf{Tr}\ln(2h)} \\
& \times \mathcal{N}_\Omega \mathcal{N}_\Phi \int d^4x \int d[\alpha] \int d[\Omega] \int d[\Xi] \int d[\Phi] \\
& \times \int d[v]\delta^{(4)}(v(s)) e^{\frac{i}{4}\int_0^s ds' [v'(s')]^2} \\
& \times e^{-i\int ds'\ \Omega^a(s')\ \alpha^a(s') - i\int ds'\ \Phi^a_{\mu\nu}(s')\Xi^a_{\mu\nu}(s')} \\
& \cdot \mathbf{tr}\left(e^{i\int_0^s ds'[\alpha^a(s') - i\sigma_{\mu\nu}\Xi^a_{\mu\nu}(s')]\tau^a}\right)_+ \\
& \times e^{-ig\int_0^s ds' v'_\mu(s')\Omega^a(s')A^a_\mu(x-\nu(s')) - 2ig\int d^4z(\partial_\nu \mathbf{\Phi}^a_{\mu\nu}(z))A^a_\mu(z)} \\
& \times e^{+ig^2\int ds'\ f^{abc}\mathbf{\Phi}^a_{\mu\nu}(s')A^b_\mu(x-v(s'))A^c_\nu(x-v(s'))} \\
& - \{g=0\},
\end{aligned} \tag{D.10}
$$

where the same properties as those of $\mathbf{G}_c[A]$ can be read off readily.

Appendix E

Effective Locality and Transverse Imprecision

Before transverse imprecision was introduced, EL had the effect of attaching to the representative symbol $[f \cdot \chi(w)]^{-1}$ of each gluon bundle, exchanged between quark and/or antiquark of respective CM coordinates y_1 and y_2, a pair of delta functions, $\delta^{(4)}(w-y_1+u(s_1))\delta^{(4)}(y_1-y_2+\bar{u}(s_2)-u(s_1))$ as used in the text, or the pair $\delta^{(4)}(w-y_1+s_1p_1)\delta^{(4)}(y_1-y_2+s_2p_2-s_1p_2)$ as used in an eikonal approximation [Fried, et al. (2010)]. For either case one finds fixed values of w_0 and w_L, and $\vec{w}_\perp = \vec{y}_{1\perp} = -\vec{y}_{2\perp}$. Then, as claimed in the text, the Halpern FI can be reduced to an ordinary set of $\int d^n\chi$ integrals. In the process, though, one makes a systematic error, of the eikonal-type, by neglecting variations of the impact parameter or, correspondingly, of momentum transfer in the core parts of the matrix element. In the context of the exact expression of the first pair of delta functions above, that *ad hoc* approximation avoided the much more complicated analysis of the transverse Fradkin's difference $u_\perp(s_1) - \bar{u}_\perp(s-2)$.

With transverse imprecision now being included, the situation changes for the better, in the sense that no such approximation need be made. But this change now requires a slightly more complicated justification of the argument which replaces Halpern's FI by a set of ordinary integrals. For the question arises if this useful simplification is also true when the $\vec{w}_\perp$ inside the $[f \cdot \chi(w)]^{-1}$ factor is itself given by $\vec{y}\,'_\perp$, and is being integrated over the $\int d^2\vec{y}\,'$ in that exponential factor, as in the discussion of the text leading to (10.16). It was there noted that the replacement of that $\vec{w}_\perp$ by $\vec{y}_{1\perp}$ or $-\vec{y}_{2\perp}$ is a reasonable approximation.

The following argument is intended to give that approximation a more detailed justification.

Consider the Halpern FI

$$\mathcal{N} \int d[\chi][\det(f\cdot\chi)]^{-\frac{1}{2}} e^{\frac{i}{2}\int d^4w\chi^2 + ig\int d^2\vec{y}'_\perp\, a(\vec{y}_{1\perp}-\vec{y}'_\perp)a(\vec{y}_{2\perp}-\vec{y}'_\perp)[f\cdot\chi(\vec{y}'_\perp)]^{-1}}, \tag{E.1}$$

where $y'_\mu = (y_0; y_L, \vec{y}'_\perp)$, the normalization is defined so that the FI of (E.1) equals 1 when $g = 0$. The dependence of color, time and longitudinal coordinate has been omitted for simplification of presentation.

As in the definition of this or any such FI, $\int d^4w\chi^2$ is understood to mean $\delta^4\Sigma_{\ell=1}^{\mathcal{N}}\chi_\ell^2$, where the subscript ℓ denotes the value of χ at the space-time point $w_{\perp\ell}$, and δ^4 corresponds to a small volume surrounding that point, which is to become arbitrarily small as N becomes arbitrarily large. As mentioned in the text, residual δ-dependence will be re-expressed in terms of physically significant quantities as a last step; but for the following argument, all the transverse coordinate differences are to be taken as arbitrarily small.

Now, re-scale the χ_ℓ variables such that $\delta^2\chi_\ell = \bar{\chi}_\ell$, and re-write (E.1) as

$$\bar{\mathcal{N}} \int d[\bar{\chi}]\; [\det(f\cdot\bar{\chi})]^{-\frac{1}{2}} e^{\frac{i}{4}\Sigma_\ell\bar{\chi}_\ell^2 + ig\delta^2\int d^2\vec{y}'_\perp a(\vec{y}_{1\perp}-\vec{y}'_\perp)a(\vec{y}_{2\perp}-\vec{y}'_\perp)[f\cdot\bar{\chi}(\vec{y}'_\perp)]^{-1}}. \tag{E.2}$$

Let us also break up the $\int d^2\vec{y}'_\perp$ integral into an infinite series of terms: one is free to choose the individual $\vec{y}'_\perp$ coordinates as exactly those which define the transverse positions of the $\bar{\chi}_\ell = \bar{\chi}(w_\ell)$. In this way, (E.2) may be re-written as

$$\bar{\mathcal{N}} \int d[\bar{\chi}]\; [\det(f\cdot\bar{\chi})]^{-\frac{1}{2}} e^{\frac{i}{4}\Sigma_\ell\bar{\chi}_\ell^2 + ig\delta^2\Delta_{y\perp}^2\; \Sigma_\ell a(\vec{y}_{1\perp}-\vec{y}'_{\perp\ell})\; a(\vec{y}_{2\perp}-\vec{y}'_{\perp\ell})[f\cdot\bar{\chi}(\vec{y}'_{\perp\ell})]^{-1}}, \tag{E.3}$$

where $\Delta_{y\perp}^2$ is understood as a true infinitesimal quantity, and where, for simplicity, we suppress explicit dependence on y_0 and y_L. But now (E.3) may be written as the product of N integrals,

$$\begin{aligned}
&\prod_\ell^N \mathcal{N}_\ell \int d^n\bar{\chi}(\vec{y}'_{\perp\ell})[\det(f\cdot\bar{\chi}(\vec{y}'_{\perp\ell})]^{-\frac{1}{2}} \\
&\quad e^{\frac{i}{4}\bar{\chi}^2(\vec{y}'_{\perp\ell}) + ig\delta^2\Delta_{y\perp}^2 a(\vec{y}_{1\perp}-\vec{y}'_{\perp\ell})a(\vec{y}_{2\perp}-\vec{y}'_{\perp\ell})[f\cdot\bar{\chi}(\vec{y}'_{\perp\ell})]^{-1}} \\
&\equiv \prod_\ell^N \mathbb{F}(ig\delta^2\Delta_{u\perp}^2\; a(\vec{y}_{1\perp}-\vec{y}_{\perp\ell})\; a(\vec{y}_{2\perp}-\vec{y}'_{\perp\ell})) \\
&\equiv \prod_\ell^N \mathbb{F}(z_\ell),
\end{aligned} \tag{E.4}$$

where (E.4) denotes the normalized product of all such $(\vec{y}'_{\perp\ell})$-valued integrals, and $\mathbb{F}(z_\ell)$ denotes the ordinary integral $\int d^n\bar{\chi}_\ell$ over the variable associated with $\vec{y}'_{\perp\ell}$. That integral is well defined in the sense that, for $|z_\ell| < 1$, as is the case here, it can be expressed as an absolutely-convergent series, or as a converging integral over a set of eigenvalues in a random matrix calculation.

One then expects to be able to write $\mathbb{F}(z_\ell)$ in terms of its Fourier transform,

$$\mathbb{F}(z_\ell) = \int d\varrho \tilde{\mathbb{F}}(\varrho) e^{iz_\ell \varrho}, \tag{E.5}$$

where the normalization condition of (E.4) stipulates that $\int d\varrho \tilde{\mathbb{F}}(\varrho) = 1$. Since z_ℓ is proportional to the infinitesimal $\delta^2_{y\perp}$ one may expand in powers of z_ℓ,

$$\mathbb{F}(z_\ell) = \int d\varrho \tilde{\mathbb{F}}(\varrho)[1 + iz_\ell\varrho + \cdots] = 1 + i\int d\varrho\ \varrho\tilde{\mathbb{F}}(\varrho) z_\ell + \cdots, \tag{E.6}$$

so that (E.4) becomes approximately

$$\begin{aligned}
&\prod_\ell \left[1 + i\int d\varrho\ \varrho\tilde{\mathbb{F}}(\varrho) z_\ell\right] \\
&= 1 + i\int d\varrho\ \varrho\tilde{\mathbb{F}}(\varrho) \sum_\ell z_\ell \\
&= 1 + i\int d\varrho\ \varrho\tilde{\mathbb{F}}(\varrho)\cdot ig\delta^2 \int d^2\vec{y}'_\perp\ a(\vec{y}_{1\perp} - \vec{y}'_\perp)\ a(\vec{y}_{2\perp} - \vec{y}'_\perp).
\end{aligned} \tag{E.7}$$

With

$$\int d^2\vec{y}'_\perp\ a(\vec{y}_{1\perp} - \vec{y}'_\perp)\ a(\vec{y}_{2\perp} - \vec{y}'_\perp) = \int \frac{d^2q}{(2\pi)^2}\ |\tilde{a}(q)|^2\ e^{iq\cdot(\vec{y}_{1\perp} - \vec{y}_{2\perp})} \equiv \varphi(\vec{b}), \tag{E.8}$$

Equation (E.4) becomes

$$\int d\varrho\tilde{\mathbb{F}}(\varrho)\ \left[1 + i\varrho(ig\delta^2)\varphi(\vec{b})\right], \tag{E.9}$$

which is just the first-order expansion of the result obtained in the text when $\vec{y}'_\perp$ was shifted to $\vec{y}'_{1\perp}$ or $-\vec{y}'_{2\perp}$. And since $g\delta^2\varphi$ is expected to be small, $\delta^2\varphi \ll 1$, it is, in effect, equivalent to

$$\int d\varrho\tilde{\mathbb{F}}(\varrho) e^{i\varrho(ig\delta^2)\varphi(\vec{b})},$$

which is just the integral of (E.2) when the intuitively equivalent change $\vec{y}'_{\perp} \to \vec{y}'_{1\perp}$ has been made in the argument of $(f \cdot \bar{\chi})^{-1}$, and after the residual δ^2 dependence has been continued to the measurably-significant value of $1/(\mu E)$. Only the first-order form, corresponding to (E.9) is needed in the calculations of quark- and nucleon-binding.

Appendix F

Tachyonic Photon Emission

A brief sketch of the probability/time "intrinsic bremsstrahlung" production, to lowest perturbative order, equation (14.32), proceeds as follows.

The S-matrix element of the lowest order, free field, photon and tachyon operators:

$$< p', k|S|p >= ie \int d^4x < p', k| : \bar{\psi}_T(x)\gamma_5\gamma \cdot A(x)\psi_T(x) : |p > \tag{F.1}$$

may be calculated by inserting into (F.1) the conventional photon operator, and the tachyon field operator (14.7) and its adjoint, which yields:

$$< p', k|S|p >= \frac{ie}{\sqrt{2\pi}}\sqrt{\frac{M^2}{EE'}}\frac{1}{\sqrt{2\omega}}(\bar{u}_{s'}(p')\gamma_5\gamma \cdot \epsilon\, u_s(p))\delta^{(4)}(p-p'-k) \tag{F.2}$$

where ϵ_μ is the photon polarization, and we use $\vec{p}$ and $E(p) = \sqrt{\vec{p}^2 - M^2}$ to represent the tachyon coordinate. Upon calculation $| < p', k|S|p > |^2$, the square of $\delta^{(4)}(p-p'-k)$ is, as always, replaced by $(2\pi)V\tau\delta^{(4)}(p-p'-k)$. The relevant quantity to calculate is then not $| < p', k|S|p > |^2$ summed over all $\vec{p}$ and $\vec{k}$ values and polarizations (and averaged over initial and summed over final spin indices), but the probability for this process to occur per unit time:

$$\frac{1}{\tau}\sum | < p', k|S|p > |^2 = \frac{M^2e^2}{(2\pi)^2E}\int \frac{d^3k}{2\omega}\frac{\delta(E(\vec{p}) - E(\vec{p}-\vec{k}) - \omega)}{E(\vec{p}) - \omega} \times \sum_{s,s'}\frac{1}{2}|\bar{u}_{s'}(p-k)\gamma_5\gamma \cdot \epsilon\, u_s(p)|^2. \tag{F.3}$$

The spin and polarization sums produce exactly unity, and the continuous energy δ function of (F.3) may be replaced by $(E-\omega)/(\omega p)\delta(\cos\theta - E/p)$

where θ is the angle — fixed by the conservation laws — between $\hat{p}$ and $\hat{k}$. The result is:

$$\frac{\alpha M^2}{pE}\int_0^{\omega_{\max}} d\omega$$

and we choose the maximum possible value of $\omega_{\max}$ as E, yielding the result quoted in Section 14.4.

Appendix G

Relativistic Tachyon Notation

In this appendix, written by Professor Y. Gabellini, the more familiar metric $(+,-,-,-)$ will be used.

G.1 Free Tachyon Equation

The γ matrices satisfy:

$$\{\gamma^\mu, \gamma^\nu\} = 2g^{\mu\nu}.$$

In the standard representation:

$$\gamma^0 = \begin{pmatrix} I & 0 \\ 0 & -I \end{pmatrix}, \quad \vec{\gamma} = \begin{pmatrix} 0 & \vec{\sigma} \\ -\vec{\sigma} & 0 \end{pmatrix}, \quad \gamma^5 = i\gamma^0\gamma^1\gamma^2\gamma^3 = \begin{pmatrix} 0 & I \\ I & 0 \end{pmatrix}.$$

In the free case, the tachyon equation is given by:

$$(i\gamma^\mu\partial_\mu - iM)\psi(x) = 0 \tag{G.1}$$

and:

$$\psi^\dagger(x)\gamma^5\gamma^0(i\gamma^\mu\overleftarrow{\partial}_\mu + iM) = 0.$$

From these two equations, a conserved current can be built:

$$j^\mu(x) = \psi^\dagger\gamma^5\gamma^0\gamma^\mu\psi(x) = j^{\mu\dagger}.$$

It follows that $j^0 = \psi^\dagger\gamma^5\psi$.

The tachyon equation can be obtained from the Lagrangian density:

$$\mathcal{L}(x) = \psi^\dagger(x)\gamma^5\gamma^0(i\gamma^\mu \overleftrightarrow{\partial}_\mu - iM)\psi(x) = \mathcal{L}^\dagger(x).$$

One gets for the conjugate momentum field π:

$$\pi(x) = \frac{\partial\mathcal{L}(x)}{\partial\dot{\psi}(x)} = i\psi^\dagger(x)\gamma^5.$$

The Hamiltonian density is then:

$$\mathcal{H}(x) = \psi^\dagger(x)\gamma^5\gamma^0[-i\vec{\gamma}\cdot\vec{\nabla} + iM]\psi(x) = i\psi^\dagger(x)\gamma^5\dot{\psi}(x) = \mathcal{H}^\dagger(x).$$

G.2 Plane Wave Solutions

G.2.1 *Positive energy solutions*

We set $\psi(x) = u(p)e^{-ip\cdot x}$. Inserted into (G.1), it gives:

$$(\not{p} - iM)u(p) = 0. \tag{G.2}$$

The spectrum of $\not{p} - iM$ being $(-2iM, -2iM, 0, 0)$ there will be two independent solutions $u^{(1)}(p)$ and $u^{(2)}(p)$ for (G.2).

They can be obtained by using the relation: $(\not{p} - iM)(\not{p} + iM) = p^2 + M^2 = 0$.

One has:

$$\not{p} + iM = E\gamma^0 - \vec{p}\cdot\vec{\gamma} + iM = \begin{pmatrix} E+iM & 0 & -p_z & -p_x+ip_y \\ 0 & E+iM & -p_x-ip_y & p_z \\ p_z & p_x-ip_y & -E+iM & 0 \\ p_x+ip_y & -p_z & 0 & -E+iM \end{pmatrix}.$$

Applying it to the two vectors $u_1 = \begin{pmatrix}1\\0\\0\\0\end{pmatrix}$ and $u_2 = \begin{pmatrix}0\\1\\0\\0\end{pmatrix}$, one obtains:

$$u^{(1)}(p) = \frac{1}{\sqrt{2iM(E+iM)}}\begin{pmatrix} E+iM \\ 0 \\ p_z \\ p_x+ip_y \end{pmatrix},$$

and

$$u^{(2)}(p) = \frac{1}{\sqrt{2iM(E+iM)}} \begin{pmatrix} p_x - ip_y \\ -p_z \\ 0 \\ E+iM \end{pmatrix}.$$

The u_1 and u_2 vectors, and the normalization of $u^{(1)}(p)$ and $u^{(2)}(p)$, have been chosen in order to mimic the standard solutions of the Dirac equation (see, for instance, [Bjorken and Drell (1965)] and [Itzykson and Zuber (1980)]), the electron mass m being simply replaced here by iM, the mass of the tachyon.

G.2.2 *Negative energy solutions*

We set $\psi(x) = v(p)e^{ip\cdot x}$. Substituted into (G.1) it gives:

$$(\not{p} + iM)v(p) = 0.$$

One notices that:

$$(\not{p} + iM)\gamma^5 u(p) = -\gamma^5(\not{p} - iM)u(p) = 0. \tag{G.3}$$

One then chooses $v^{(1)}(p) = \gamma^5 u^{(1)}(p)$, $v^{(2)}(p) = \gamma^5 u^{(2)}(p)$, and obtains:

$$v^{(1)}(p) = \frac{1}{\sqrt{2iM(E+iM)}} \begin{pmatrix} p_z \\ p_x + ip_y \\ E+iM \\ 0 \end{pmatrix},$$

and

$$v^{(2)}(p) = \frac{1}{\sqrt{2iM(E+iM)}} \begin{pmatrix} p_x - ip_y \\ -p_z \\ 0 \\ E+iM \end{pmatrix}.$$

As for $u^{(1)}(p)$ and $u^{(2)}(p)$, the vectors $v^{(1)}(p)$ and $v^{(2)}(p)$ have been chosen in order to mimic the standard negative energy solutions of the Dirac equation.

G.3 Some Useful Formulae, Valid for $|\vec{p}| \geq M$

$$
\begin{aligned}
&u^{\dagger(\alpha)}(p)u^{(\beta)}(p) = \frac{|\vec{p}|}{M} \\
&u^{\dagger(\alpha)}(p)\gamma^0 u^{(\beta)}(p) = 0 \\
&u^{\dagger(\alpha)}(p)\gamma^5 u^{(\beta)}(p) = \frac{E}{iM} i(\vec{\sigma}\cdot\hat{p})_{\alpha\beta} \\
&u^{\dagger(\alpha)}(p)\gamma^5 u^{(\beta)}(p)(-i\vec{\sigma}\cdot\hat{p})_{\beta\alpha'} = \frac{E}{iM}\delta_{\alpha\alpha'} \\
&u^{\dagger(\alpha)}(p)\gamma^5\gamma^0 u^{(\beta)}(p) = i(\vec{\sigma}\cdot\hat{p})_{\alpha\beta} \\
&u^{\dagger(\alpha)}(p)\gamma^5\gamma^0 u^{(\beta)}(p)(-i\vec{\sigma}\cdot\hat{p})_{\beta\alpha'} = \delta_{\alpha\alpha'} \\
&u^{(\alpha)}(p)(-i\vec{\sigma}\cdot\hat{p})_{\alpha\beta} \bigotimes u^{\dagger(\beta)}(p)\gamma^5\gamma^0 = \frac{\not{p} + iM}{2iM} \\
&v^{\dagger(\alpha)}(p)v^{(\beta)}(p) = \frac{|\vec{p}|}{M} \\
&v^{\dagger(\alpha)}(p)\gamma^0 v^{(\beta)}(p) = 0 \\
&v^{\dagger(\alpha)}(p)\gamma^5 v^{(\beta)}(p) = \frac{E}{iM} i(\vec{\sigma}\cdot\hat{p})_{\alpha\beta} \\
&v^{\dagger(\alpha)}(p)\gamma^5 v^{(\beta)}(p)(-i\vec{\sigma}\cdot\hat{p})_{\beta\alpha'} = \frac{E}{iM}\delta_{\alpha\alpha'} \\
&v^{\dagger(\alpha)}(p)\gamma^5\gamma^0 v^{(\beta)}(p) = -i(\vec{\sigma}\cdot\hat{p})_{\alpha\beta} \\
&v^{\dagger(\alpha)}(p)\gamma^5\gamma^0 v^{(\beta)}(p)(-i\vec{\sigma}\cdot\hat{p})_{\beta\alpha'} = -\delta_{\alpha\alpha'} \\
&v^{(\alpha)}(p)(-i\vec{\sigma}\cdot\hat{p})_{\alpha\beta} \otimes v^{\dagger(\beta)}(p)\gamma^5\gamma^0 = \frac{\not{p} - iM}{2iM} \\
&v^{\dagger(\alpha)}(E,-\vec{p})\gamma^5 u^{(\beta)}(E,\vec{p}) = 0 \\
&u^{\dagger(\alpha)}(E,-\vec{p})\gamma^5 v^{(\beta)}(E,\vec{p}) = 0 \\
&v^{\dagger(\alpha)}(p)\gamma^5\gamma^0 u^{(\beta)}(p) = 0 \\
&u^{\dagger(\alpha)}(p)\gamma^5\gamma^0 v^{(\beta)}(p) = 0.
\end{aligned}
$$

G.4 "Spin" of the Tachyon

The Lorentz transform generators leaving the tachyon equation invariant are:

$$\sum\nolimits_{\mu\nu} = -\frac{i}{4}[\gamma_\mu, \gamma_\nu].$$

Using the standard representations, one finds:

$$\sum_{0i} = -\frac{i}{2}\begin{pmatrix} 0 & \sigma_i \\ \sigma_i & 0 \end{pmatrix}, \qquad \sum_{ij} = -\frac{1}{2}\epsilon_{ijk}\begin{pmatrix} \sigma_k & 0 \\ 0 & \sigma_k \end{pmatrix}.$$

One then obtains the spin of the tachyon by looking at the Pauli–Lubanski operator:

$$W_\mu = \frac{1}{2}\epsilon_{\mu\nu\rho\sigma}P^\nu J^{\rho\sigma}$$

with:

$$J_{\alpha\beta} = i(x_\alpha \partial_\beta - x_\beta \partial_\alpha) + \Sigma_{\alpha\beta}.$$

One then has:

$$W_\mu = \frac{1}{2}\epsilon_{\mu\nu\rho\sigma}P^\nu \Sigma^{\rho\sigma}.$$

By choosing the characteristic vector: $P_\mu = (0,0,0,M)$, one obtains immediately:

$$W_\mu = M(\Sigma^{12}, \Sigma^{20}, \Sigma^{01}, 0).$$

The first three components of W_μ are nothing by the generator of the SU(1,1) Lie group. And one finds:

$$W_\mu W^\mu = M^2({\Sigma^{12}}^2 - {\Sigma^{20}}^2 - {\Sigma^{01}}^2) = \frac{3}{4}M^2. \tag{G.4}$$

From the relation $W_\mu W^\mu = -M^2 s(s+1)$, one extracts: $s = -\frac{1}{2} \pm \frac{i}{\sqrt{2}}$.

G.5 Quantization

We write the solution of (G.1) in the form:

$$\psi(x) = \int \frac{d^3p}{(2\pi)^3}\frac{M}{E}\theta(|\vec{p}| - M) \sum_{\alpha,\beta=1}^{2} [u^{(\alpha)}(p)(\vec{\sigma}\cdot\hat{p})_{\alpha\beta} b_\beta(\vec{p}) e^{-ip\cdot x}$$
$$+ v^{(\alpha)}(p)(\vec{\sigma}\cdot\hat{p})_{\alpha\beta} d^\dagger_\beta(\vec{p}) e^{ip\cdot x}]. \tag{G.5}$$

The measure $\tilde{dp} = \frac{d^3p}{(2\pi)^3}\frac{M}{E}\theta(|\vec{p}| - M)$ with $E = \sqrt{\vec{p}^2 - M^2}$ is Lorentz invariant.

One quantizes according to:

$$\{b_\alpha(\vec{p}), b^\dagger_\beta(\vec{p})'\} = (2\pi)^3 \frac{E}{M}\delta^3(\vec{p} - \vec{p}')(\vec{\sigma}\cdot\hat{p})_{\alpha\beta}$$
$$\{d_\alpha(\vec{p}), d^\dagger_\beta(\vec{p}')\} = (2\pi)^3 \frac{E}{M}\delta^3(\vec{p} - \vec{p}')(\vec{\sigma}\cdot\hat{p})_{\beta\alpha}.$$

All the other anti-commutators vanish.
One then obtains:

$$\begin{aligned}\{\psi_i(\vec{x},t), \pi_j(\vec{y},t)\} &= \{\psi_i(\vec{x},t), (i\psi^\dagger(\vec{y},t)\gamma^5)_j\} \\ &= i\delta_{ij}\int \frac{d^3p}{(2\pi)^3}\theta(|\vec{p}| - M)e^{i\vec{p}\cdot(\vec{x}-\vec{y})}.\end{aligned}$$

Or, using (14.16):

$$\{\psi_i(\vec{x},t), \pi_j(\vec{y},t)\} = i\delta_{ij}\hat{\delta}^{(3)}(\vec{x} - \vec{y}). \tag{G.6}$$

The Hamiltonian operator is then:

$$H = \int d^3x\ \mathcal{H}(x) = \int \tilde{dp}E\sum_{\alpha,\beta}\left[b^\dagger_\alpha(\vec{p})b_\beta(\vec{p}) - d_\alpha(\vec{p})d^\dagger_\beta(\vec{p})\right](\vec{\sigma}\cdot\hat{p})_{\alpha\beta}.$$

As a generalization, one has:

$$\begin{aligned}P^\mu &= \int d^3x\ \psi^\dagger(x)\gamma^5 i\partial^\mu\psi(x) \\ &= \int \tilde{dp}\ p^\mu\sum_{\alpha,\beta}\left[b^\dagger_\alpha(\vec{p})b_\beta(\vec{p}) - d_\alpha(\vec{p})d^\dagger_\beta(\vec{p})\right](\vec{\sigma}\cdot\hat{p})_{\alpha\beta}.\end{aligned}$$

In the normal form, it gives:

$$P^\mu = \int \tilde{dp}\ p^\mu\sum_{\alpha,\beta}\left[b^\dagger_\alpha(\vec{p})b_\beta(\vec{p}) + d^\dagger_\beta(\vec{p})d_\alpha(\vec{p})\right](\vec{\sigma}\cdot\hat{p})_{\alpha\beta}. \tag{G.7}$$

And, as $(\vec{\sigma}\cdot\hat{p})^*_{\alpha\beta} = (\vec{\sigma}\cdot\hat{p})_{\beta\alpha}$, one gets, as expected: $P^\mu = P^{\dagger\mu}$.
We obtain the usual commutation relations:

$$[P^\mu, b_\alpha(\vec{p})] = -p^\mu b_\alpha(\vec{p}), \qquad [P^\mu, b^\dagger_\alpha(\vec{p})] = p^\mu b^\dagger_\alpha(\vec{p})$$
$$[P^\mu, d_\alpha(\vec{p})] = -p^\mu d_\alpha(\vec{p}), \qquad [P^\mu, d^\dagger_\alpha(\vec{p})] = p^\mu d^\dagger_\alpha(\vec{p}).$$

The charge operator is:

$$Q = \int d^3x\ j^0(x) = \int d^3x\ \psi^\dagger(x)\gamma^5\psi(x)$$
$$= \int \tilde{dp} \sum_{\alpha,\beta} \left[b^\dagger_\alpha(\vec{p})b_\beta(\vec{p}) - d^\dagger_\beta(\vec{p})d_\alpha(\vec{p})\right](\vec{\sigma}\cdot\hat{p})_{\alpha\beta}.$$

It follows that:

$$[Q, b_\alpha(\vec{p})] = -b_\alpha(\vec{p}), \qquad [Q, b^\dagger_\alpha(\vec{p})] = b^\dagger_\alpha(\vec{p})$$
$$[Q, d_\alpha(\vec{p})] = d_\alpha(\vec{p}), \qquad [Q, d^\dagger_\alpha(\vec{p})] = -d^\dagger_\alpha(\vec{p}).$$

G.6 Tachyonic States

We define:

$$b_\alpha(\vec{p})|0>= 0, \qquad d_\alpha(\vec{p})|0>= 0$$
$$b^\dagger_\alpha(\vec{p})|0>= |\vec{p},\alpha>_T, \qquad d^\dagger_\alpha(\vec{p})|0>= |\vec{p},\alpha>_{\bar{T}}\ .$$

Then:

$$< 0|\psi(x)|\vec{p},\alpha>_T = u^{(\alpha)}(p)e^{-ip\cdot x}, \quad {}_{\bar{T}}<\vec{p},\alpha|\psi(x)|0>= v^{(\alpha)}(p)e^{ip\cdot x}. \tag{G.8}$$

One finds the action of the quadri-momentum operator:

$$P^\mu|\vec{q},\gamma>_T = \int \tilde{dp}\ p^\mu \sum_{\alpha,\beta}(\vec{\sigma}\cdot\hat{p})_{\alpha\beta}\left[b^\dagger_\alpha(\vec{p})b_\beta(\vec{p})\right.$$
$$\left. + d^\dagger_\beta(\vec{p})d_\alpha(\vec{p})\right]b^\dagger_\gamma(\vec{q})|0>$$
$$= \int \tilde{dp}\ p^\mu \sum_{\alpha,\beta}(\vec{\sigma}\cdot\hat{p})_{\alpha,\beta}b^\dagger_\alpha(\vec{p})(2\pi)^3\frac{E}{M}\delta^3(\vec{p}-\vec{q})(\vec{\sigma}\cdot\hat{p})_{\beta\gamma}|0>$$
$$= q^\mu b^\dagger_\gamma(\vec{q})|0>= q^\mu|\vec{q},\gamma>_T$$

The state vectors are normalized according to:

$$< 0|0 > = 1$$
$${}_T<\vec{p},\alpha|\vec{q},\beta>_T = (2\pi)^3\frac{E}{M}\delta^3(\vec{p}-\vec{q})(\vec{\sigma}\cdot\hat{p})_{\alpha\beta}$$
$${}_{\bar{T}}<\vec{p},\alpha|\vec{q},\beta>_{\bar{T}} = (2\pi)^3\frac{E}{M}\delta^3(\vec{p}-\vec{q})(\vec{\sigma}\cdot\hat{p})_{\beta\alpha}.$$

First Epilogue

Once upon a time, at the Pearly Gates in the not-too-distant future, the following conversations might take place.

St. Peter: Professor Feynman, what are you doing here? You were admitted some time ago.

Feynman: I have received information that this fellow Fried will be arriving shortly; and when he does appear, I want to give him a piece of my mind!

St. Peter: Why? Has he done something wrong that I don't know about?

Feynman: Wrong? You damned right he has! He's been mocking my Graphs, saying that they're useless, while pushing some cockamamy functional formalism!

A sound of wings and a new voice is heard on the other side of St. Peter.

Schwinger: Now, now Richard, aren't you being a bit hasty? Fried never scorned your Graphs; rather, he insisted that a sum of all such relevant Graphs can only be done functionally.

Feynman: Oh, come on Julian! He's been bad-mouthing me in that last book of his, and it's getting on my nerves!

Another flurry of wings, and a new voice joins the conversation.

Fradkin: You've got it all wrong, Feynman! It's Julian who might be taking offense, not you, because without my Representations his pretty solutions, while correct, would be useless for non-perturbative problems!

Schwinger: I most certainly do not take offense at Fried's position, Efimov, because I quite agree: Your Representations do permit my functional solutions to be applied to non-perturbative questions. Only, as I'm sure you'll agree, that while your L[A] Representation allowed the k=0 sum over Richard's Graphs to be estimated in the calculation of charge renormalization, that statement cannot be made for non-zero momenta of the photon propagator.

Fradkin: Da, you have a point. But for any perturbative order of approximation, the use of my L[A] representation simplifies the problem tremendously.

Feynman: Well, whatever you guys say, I'm still furious. Functional field theory: Ha!

A loud flurry of wings overhead is heard, a murmur of several voices shouts: "Humbug! Humbug!"; and a raucous voice is heard to say: " Humbug! The less they know about Field Theory, the better off they'll be!" And then all is silent. The three Professors turn to St. Peter for an explanation, who shrugs his shoulders and says:

St. Peter: "Oh, pay them no attention. They're just some unreconciled S-Matrix theoreticians who cannot accept defeat."

Second Epilogue

Two theoretical physicists, Able and Baker, had been sitting next to each other on a small, commuting aircraft, on their way to lecture at a nearby university, when an explosion cut short their flight and their lives. But they awoke, seemingly intact, although in strange clothes of tight-fitting plastic, on chairs in a small room. The room was extremely hot, and they gasped for breath, as they spoke,

Able: What, how, where are we?

Baker: All I remember is that explosion; we must have been killed!

A panel of the wall opposite them slid open, and an orange-clad Fiend entered, smiling as he stood before them in clothes impervious to the rivulets of fire which ran up and down his sleeves and trousers, and circled about his body. There was fire in his eyes as he spoke to them, but in tones that were almost musical,

Fiend: Good evening, gentlemen. I see you have arrived safely.

Baker: Who, what, where are we? And why is it so damned hot?

Fiend: Surely you must have guessed by now. You are in Hell, gentlemen, or at least a small suburb of it. And here you shall stay for a short while, in penance for the particular sin of which you both are guilty.

Able: Sin? What sin? What are you talking about? I'll have to let you know I'm a good Unitarian; I don't commit sins!

Fiend: Gentlemen, please. I know your histories, and I can read your thoughts. The sin of which you both are guilty, although for different reasons, and for which you will here do penance, is a sin against the very structure of your chosen profession.

Here the Fiend chuckled, but immediately became serious again.

When acting as Referees, you damned an unknown colleague's work: You, Able, because if his result was correct, a previous result of yours would necessarily be incorrect; and you could not face that. You, Baker, because you were just lazy, or should I say, too damned lazy?

And here the Fiend chuckled again.

Baker: Now see here, You, whoever you are. We are, or perhaps were, both very distinguished physicists! You can't just treat us like this! And if You're referring to the Review I gave in which the author's conclusion violated a well-founded and accepted Theory, how can you object to my Report?

Fiend: Ah, but Yes I can; and I shall answer your statement by quoting the words of a favorite poet[4] which express the sentiment that should be in the heart of every true physicist: "There's a crack in everything; that's how the light gets in."

Able: On a more practical note, just long are You going to keep us here, in this damned hot box?

Fiend: Oh, not too long...

And glancing at his wristwatch, he added:

Fiend: Oh, I should say, not more than three thousand years. But now, I must bids you Good Day.

And taking one step backwards, the Fiend vanished, disappearing into the fifth dimension.

[4]Leonard Cohen, from his poem "Anthem".

Bibliography

Andersen H.C., Chandler D., and Weeks J.D. (1976) *Adv. Chem. Phys*, 34, p.105.

Avan J., Fried H.M. and Gabellini Y. (2003) *Phys. Rev.* D67, 16003.

Baker M. and Johnson K. (1969). *Phys. Rev.*, 183, p. 1292.

Balibar F., Laverne A. and Levy Leblond J.M. (1989). *Quantique: Elements* (North Holland Press).

Baym G. (1960) *Phys. Rev.* 117, p.886.

Beckers J. and Jaspers M. (1978) *Ann. Phys.* 113, No.2, 237.

Bender C.M., et al. (1979) *Phys. Rev.* D20, p. 1374.

Bender C.M., et al. (1988) *Phys. Rev.* D37, p.1472.

Berezin F.A. (1961) *Dokl. Akad. Nauk SSSR*, 137, p.311.

Bjorken J.D. and Drell S. (1965) *Relativistic Quantum Fields* (McGraw-Hill).

Bloch F. and Nordsieck A. (1937)*Phy. Rev.* 52, p. 54.

Bogoluibov N.N. and Shirkov D.V. (1976) *Introduction to the Theory of Quantized Fields* (J. Wiley & Sons).

Brachet M.E. and Fried H.M. (1984) *Phys. Letters* 103A, p.309.

Brachet M.E. and Fried H.M. (1987) *J. Math. Phys* 28, p.15.

Daigne F. *Gamma Ray Bursts*, EDP Sciences, Springer-Verlag, 1999.

Desjardins S.G. and Stratt R.M. (1984). *J.Chem. Phys.* 81, p.6232.

Dhar J. and Sudarshan E.C.G. (1968)*Phys. Rev.* 174, No.5, 1808.

Dirac, P.A.M. (1943) *Comm. Dublin Inst. Adv. Studies A.* **1**

Dirac P.A.M. (1958) *Proc. Roy. Soc.* A246, p.326.

Faddeev L.D. and Slavnov A.A. (1980) *Gauge Fields - Introduction to Quantum Field Theory* (Benjamin Cummings).

Feynman R.P.(1948) *Rev. Mod. Phys.* 20, p. 367.

Feynman R.P.(1949) *Phys. Rev.* 76, (Theory of Positrons, p. 749) and (Space-Time Approach to QED p. 769).

Feynman R.P.(1962) *Quantum Electrodynamics* (W.A. Benjamin).

Feynman R.P. and Hibbs A.R. (1965) *Quantum Mechanics and Path Integrals* (McGraw-Hill).

Finkbeiner D., Su M. and Slatyer T. (2010) *The Astrophysical Journal*, November 10.

Frontera F. (2004) arXiv: astro-ph/0407633.

Fradkin E.S. (1954) *Dokl. Acad. Nauk. SSSR* 98, p.47.
Fradkin E.S. (1966) *Nuc. Phys.*, 76, p.588.
Fried, H.M. (1972) *Functional Methods and Models in Quantum Field Theory* (MIT Press).
Fried H.M., Gabellini Y., Grandou T. and Sheu Y-M. (1975) *Eur. Phys. J.* C65, 395.
Fried H.M. (1980) *Nucl. Phys.* B169, p.329.
Fried H.M. (1983) *Phys. Rev.* D27, p.2956.
Fried H.M., Kang K. and McKellar B.H.J. (1983) *Phys. Rev.* A28, 738.
Fried H.M. (1987) *J. Math. Phys.* 28, p.1275.
Fried H.M. (1990) *Basics of Functional Methods and Eikonal Models* (Editions Frontieres, Gif-sur-Yvette, France).
Fried H.M. (2002) *Green's Functions and Ordered Exponentials*(Cambridge University Press).
Fried H.M. and Woodard R.P. (2002) *Phys. Lett*, B524, 233.
Fried H.M. (2003) arXiv:0310095, v.1.
Fried H.M. and Gabellini Y. (2009) *Phys. Rev.* D79, 065035.
Fried H.M. and Gabellini Y. (2007) arXiv:0709.0414, v1.
Fried H.M., Gabellini Y., Grandou T., Sheu Y-M. (2010) *Euro. Phys. J. C* 65, p.395-411.
Fried H.M., Gabellini Y., Grandou T. and Sheu Y-M. (2011) arXiv:1104.4663 [hep-th].
Fried H.M., Grandou T. and SHeu Y-M. (2012) *Annals of Physics* 327, p.2666-2690.
Fried H.M., Gabellini Y., Grandou T. and Sheu Y-M. (2012) arXiv:1203.6137, v1.
Fried H.M., Grandou T. and Sheu Y-M. work in completion
Fried H.M., Gattobigio M., Grandou T. and Sheu Y-M. arXiv:1103.2936 [hep-th].
Gell-Mann M. and Low F. (1954). *Phys. Rev.*, 95, p. 1300.
Grandou T. (2011) *On Some Aspects of the QCD Effective Locality* Proceedings of the 11th Workshop on Non-Perturbative QCD, Paris (edited by Muller B. and Tan C.I.)
Greisen K. (1966) *Phys. Rev. Lett.* 16, 748.
Gribov V.N. (1978) *Nucl. Phys.* B139, p.1
Guay A. (2004) *Geometrical aspects of local gauge symmetry.*
Grandou T., Sheu Y-M. and Fried H.M. (2013) INLN Report *Non-Perturbative QCD Amplitudes in Quenched and Eikonal Approximation.*
Guralnik G.S. (1964) *Phys.Rev.* 136, 1401.
Halpern M.B. (1977)*Phys. Rev.* D16, 1798.
Halpern M.B. (1977)*Phys. Rev.* D16, 3515.
Heitler W. *The Quantum Theory of Radiation*, Oxford, 1954.
Hofmann R. (2006) *Int. J. Mod. Phys.* A20, 4123; *Erratum-ibid* A21, 6515.
Hounkonnou M.N. and Naciri M. (2000) *J. Phys.* G26, 1849.
Islam M. (2011) *Proton structure and prediction of pp elastic scattering of 7 TeV*, Proceedings of the 11th Workshop on Non-Perturbative QCD, Paris, France, June 2011.

Itzykson C. and Zuber J-B. (1980) *Quantum Field Theory* (McGraw-Hill).
Jastrow R. (1951). *Phys. Rev.* 81, 664.
Johnson K., Wiley R. and Baker M. (1967). *Phys. Rev.*, 167, p. 1699.
Jost R. and Luttinger J.M. (1950). *Phys. Acta*, 23, p. 201.
Kalle G. and Dan K.(1953) *Vidensk. Selsk.*, 27, 12. Reproduced in *Selected Papers on Quantum Electrodynamics,* Dover (1958), edited by J. Schwinger.
Kirshner R.P. (1999) *Proc. Natl. Acad. Sci. USA*96, 4224.
Levy M. (1964) *Nucl. Phys.* 57, p. 152.
Liddle A.R. and Lyth D.H. (2000) *Cosmological Inflation and Large-Scale Structure,* Cambridge University Press, UK.
Luscher M., Symanzik K. and Weisz P. (1980) *Nucl. Phys.* B173, 365.
Luscher M. (1981) *Nucl. Phys.* B180, 317.
Mitter P.K. (1973) Proceedings of the Cargese Lectures.
Nambu Y. (1979) *Phys.Lett.* B80,372.
Piran T. (2004) *Rev. Mod. Phys.* 76, 1143.
Rafelski J., Labun L., Hadad Y. and Chen P. (2009) arXiv:0909.2989 [gr-qc].
Reinhardt H., Langfeld K and Smekal L.V. (1993) *Phys.Lett.* B300, 111.
Reiss A.G., et al. (2000) *The Astrophysical Journal.* 536, 62-67.
Rzewuski, J. (1972) *Field Theory* (PWN Publishers).
Schwinger, J. (1949) *Phys. Rev.*, 76, p. 790.
Schwinger, J. (1951) *Proc. Nat'l Acad. Sciences*, 37, p. 452.
Schwinger J. (1951) *Phys. Rev.* 82, p.664.
Schwinger, J. (1954) Harvard Lectures.
Schwinger, J. (1956) Stanford Lectures.
Schwinger, J. (1970) *Particles, Sources and Fields* (Addison-Wesley).
Sheu Y-M. (2008) *Finite-Temperature Quantum Electrodynamics: General Theory and Bloch-Nordsieck Estimates of Fermion Damping in a Hot Medium* PhD Thesis, Brown University, Providence, RI.
Sommerfield C. (1963). *Ann. of Phys.*, 26, p.1.
Stratt R.M. (1984) *J. Chem. Phys.* 80, p.5764.
Svidinsky A.V. and Eksper Z. (1956) *Teoretich. Fiz.* 31, p. 324.
Symanzik K. (1954) *Naturforschung Z.* 92, p.809.
Symanzik K. (1960) Lectures UCLA.
Symanzik K. (1960) Private Communication, UCLA.
Symanzik K. (1961) Lectures at the Summer School for High Energy Physics at Hercegnovi, Yugoslavia.
Symanzik K. (1979) Cargese Lectures (Plenum Press).
't Hooft G. and Veltman M. (1973) *DIAGRAMMAR* (CERN Lab I Publication).
't Hooft G. (1979) *Nucl. Phys.* B153, p.141.
Tomaras T.N., Tsamis N.C. and Woodard, R.P. (2000) *Phys. Rev.* D62, 125005.
Valatin J. (1964) *Proc. Roy. Soc.* 225A, p.535; 226A, p. 254.
Veltman M. (1975) Proceedings of the International Summer School on Particle Physics, Basko Polje, Yugoslavia.
Wick G.C. (1950) *Phys. Rev.* 80, p.268.
Wigner E.P. (1939) *Ann. of Math.* 40 ,149-204.
Yang C.N. and Feldman D. (1960) *Phys. Rev.* 79, p.972.

Zatsepin G.T. and Kuzmin V.A.(1966) *JETP Lett.* 16, 78.
Zee A. *Quantum Field Theory in a Nutshell,* Princeton University Press.
Zimmerman W. (1967) *Comm. Math. Phys.* 6, p.161.
Zimmerman W. (1968) *Comm. Math. Phys.* 10, p.325.
Zumino B. (1958). New York University Lecture Notes.
Zumino B. (1960) *J. Math. Phys.* 1, p.1
Zumino B. (1975) *Nucl. Phys.*, B89, p.535.
Zwanziger D. (1985) Proceedings of the VI Erice International School.

Printed in the United States
By Bookmasters

Printed in the United States
By Bookmasters